ONDON MATHEMATICAL SOCIETY LECTURE NOTE SERIES

anaging Editor: Professor J.W.S. Cassels, Department of Pure Mathematics and Mathematical Statistics,
iversity of Cambridge, 16 Mill Lane, Cambridge CB2 1SB, England

e books in the series listed below are available from booksellers, or, in case of difficulty,
m Cambridge University Press.

4 Algebraic topology, J.F. ADAMS
7 Differential germs and catastrophes, Th. BROCKER & L. LANDER
7 Skew field constructions, P.M. COHN
4 Representation theory of Lie groups, M.F. ATIYAH *et al*
6 Homological group theory, C.T.C. WALL (ed)
9 Affine sets and affine groups, D.G. NORTHCOTT
0 Introduction to H_p spaces, P.J. KOOSIS
2 Topics in the theory of group presentations, D.L. JOHNSON
3 Graphs, codes and designs, P.J. CAMERON & J.H. VAN LINT
5 Recursion theory: its generalisations and applications, F.R. DRAKE & S.S. WAINER (eds)
6 p-adic analysis: a short course on recent work, N. KOBLITZ
9 Finite geometries and designs, P. CAMERON, J.W.P. HIRSCHFELD & D.R. HUGHES (eds)
0 Commutator calculus and groups of homotopy classes, H.J. BAUES
1 Synthetic differential geometry, A. KOCK
4 Markov processes and related problems of analysis, E.B. DYNKIN
7 Techniques of geometric topology, R.A. FENN
8 Singularities of smooth functions and maps, J.A. MARTINET
9 Applicable differential geometry, M. CRAMPIN & F.A.E. PIRANI
0 Integrable systems, S.P. NOVIKOV *et al*
2 Economics for mathematicians, J.W.S. CASSELS
5 Several complex variables and complex manifolds I, M.J. FIELD
6 Several complex variables and complex manifolds II, M.J. FIELD
8 Complex algebraic surfaces, A. BEAUVILLE
9 Representation theory, I.M. GELFAND *et al*
4 Symmetric designs: an algebraic approach, E.S. LANDER
6 Spectral theory of linear differential operators and comparison algebras, H.O. CORDES
7 Isolated singular points on complete intersections, E.J.N. LOOIJENGA
8 A primer on Riemann surfaces, A.F. BEARDON
9 Probability, statistics and analysis, J.F.C. KINGMAN & G.E.H. REUTER (eds)
0 Introduction to the representation theory of compact and locally compact groups, A. ROBERT
1 Skew fields, P.K. DRAXL
2 Surveys in combinatorics, E.K. LLOYD (ed)
3 Homogeneous structures on Riemannian manifolds, F. TRICERRI & L. VANHECKE
4 Finite group algebras and their modules, P. LANDROCK
5 Solitons, P.G. DRAZIN
6 Topological topics, I.M. JAMES (ed)
7 Surveys in set theory, A.R.D. MATHIAS (ed)
8 FPF ring theory, C. FAITH & S. PAGE
9 An F-space sampler, N.J. KALTON, N.T. PECK & J.W. ROBERTS
0 Polytopes and symmetry, S.A. ROBERTSON
1 Classgroups of group rings, M.J. TAYLOR
2 Representation of rings over skew fields, A.H. SCHOFIELD
3 Aspects of topology, I.M. JAMES & E.H. KRONHEIMER (eds)
4 Representations of general linear groups, G.D. JAMES
5 Low-dimensional topology 1982, R.A. FENN (ed)

96 Diophantine equations over function fields, R.C. MASON
97 Varieties of constructive mathematics, D.S. BRIDGES & F. RICHMAN
98 Localization in Noetherian rings, A.V. JATEGAONKAR
99 Methods of differential geometry in algebraic topology, M. KAROUBI & C. LERUSTE
100 Stopping time techniques for analysts and probabilists, L. EGGHE
101 Groups and geometry, ROGER C. LYNDON
103 Surveys in combinatorics 1985, I. ANDERSON (ed)
104 Elliptic structures on 3-manifolds, C.B. THOMAS
105 A local spectral theory for closed operators, I. ERDELYI & WANG SHENGWANG
106 Syzygies, E.G. EVANS & P. GRIFFITH
107 Compactification of Siegel moduli schemes, C-L. CHAI
108 Some topics in graph theory, H.P. YAP
109 Diophantine Analysis, J. LOXTON & A. VAN DER POORTEN (eds)
110 An introduction to surreal numbers, H. GONSHOR
111 Analytical and geometric aspects of hyperbolic space, D.B.A.EPSTEIN (ed)
112 Low-dimensional topology and Kleinian groups, D.B.A. EPSTEIN (ed)
113 Lectures on the asymptotic theory of ideals, D. REES
114 Lectures on Bochner-Riesz means, K.M. DAVIS & Y-C. CHANG
115 An introduction to independence for analysts, H.G. DALES & W.H. WOODIN
116 Representations of algebras, P.J. WEBB (ed)
117 Homotopy theory, E. REES & J.D.S. JONES (eds)
118 Skew linear groups, M. SHIRVANI & B. WEHRFRITZ
119 Triangulated categories in the representation theory of finite-dimensional algebras, D. HAPPEL
120 Lectures on Fermat varieties, T. SHIODA
121 Proceedings of *Groups - St Andrews 1985*, E. ROBERTSON & C. CAMPBELL (eds)
122 Non-classical continuum mechanics, R.J. KNOPS & A.A. LACEY (eds)
123 Surveys in combinatorics 1987, C. WHITEHEAD (ed)
124 Lie groupoids and Lie algebroids in differential geometry, K. MACKENZIE
125 Commutator theory for congruence modular varieties, R. FREESE & R. MCKENZIE
126 Van der Corput's method for exponential sums, S.W. GRAHAM & G. KOLESNIK
127 New directions in dynamical systems, T.J. BEDFORD & J.W. SWIFT (eds)
128 Descriptive set theory and the structure of sets of uniqueness, A.S. KECHRIS & A. LOUVEAU
129 The subgroup structure of the finite classical groups, P.B. KLEIDMAN & M.W.LIEBECK
130 Model theory and modules, M. PREST
131 Algebraic, extremal & metric combinatorics, M.M. DEZA, P. FRANKL & I.G. ROSENBERG (eds)
132 Whitehead groups of finite groups, ROBERT OLIVER
133 Linear algebraic monoids, MOHAN S. PUTCHA
134 Number theory and dynamical systems, M. DODSON & J. VICKERS (eds)
135 Operator algebras and applications, 1, D. EVANS & M. TAKESAKI (eds)
136 Operator algebras and applications, 2, D. EVANS & M. TAKESAKI (eds)
137 Analysis at Urbana, I, E. BERKSON, T. PECK, & J. UHL (eds)
138 Analysis at Urbana, II, E. BERKSON, T. PECK, & J. UHL (eds)
139 Advances in homotopy theory, B. STEER & W. SUTHERLAND (eds)
140 Geometric aspects of Banach spaces E.M.PEINADOR and A.R USAN(eds)
141 Surveys in combinatorics 1989, J. SIEMONS (ed)
142 The geometry of jet bundles, D.J. SAUNDERS

London Mathematical Society Lecture Note Series. 140

Geometric Aspects of Banach Spaces

Essays in honour of Antonio Plans

Edited by
E.Martin-Peinador, Facultad de Ciencias Matematicas, Universidad Complutense Madrid
and
A.Rodés, Facultad de Ciencias, Universidad de Zaragoza

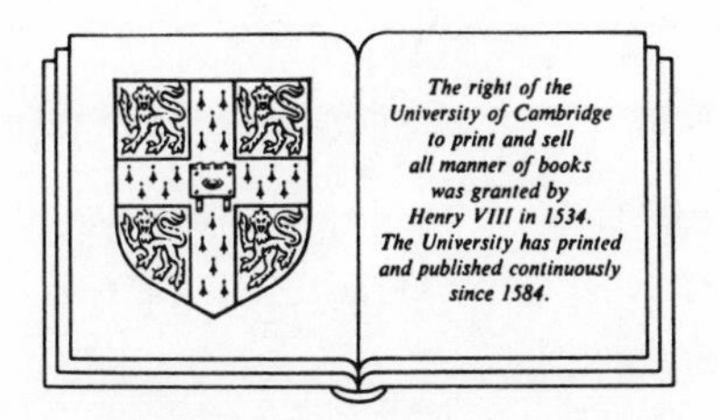

CAMBRIDGE UNIVERSITY PRESS
Cambridge
New York Port Chester Melbourne Sydney

Published by the Press Syndicate of the University of Cambridge
The Pitt Building, Trumpington Street, Cambridge CB2 1RP
32 East 57th Street, New York, NY 10022, USA
10, Stamford Road, Oakleigh, Melbourne 3166, Australia

© Cambridge University Press 1989

First published 1989

Printed in Great Britain at the University Press, Cambridge

Library of Congress cataloging in publication data available

British Library cataloguing in publication data available

ISBN 0 521 36752 2

CB-STK
SCIMON

CONTENTS.

This book has been supported by the University of Zaragoza. We are indebted to its rector, Professor Camarena, for his encouragement.

We would like to mention here Víctor Onieva, one of the first students of Professor Plans, who recently died, as this book was in preparation. His hard work and tremendous strength in the dayly struggle with his weak health will undoubtfully be rewarded as he deserved.

Thanks are due also to Pedro Ortega for his careful typing of the book.

ANTONIO PLANS: A BIOGRAPHICAL OUTLINE

The collection of papers contained in this book is intended to be a warm homage to Antonio Plans, on the occasion of his 65 th. birthday, which at present in Spain marks the point of retirement from academic undergraduate teaching.

We have chosen the topic of Banach Spaces since it has been the center of his Mathematical interests for the last few years. The papers included have been written by friends, or students of Professor Plans, or by a few of the mathematicians who have been in touch with him for scientific reasons.

We feel sorry to restrict ourselves to just one topic, since there are many colleagues who, for sure, would have liked to write something for this event. As a matter of fact Professor Plans is widely known by his results on Knot Theory, a field he worked on at the beginning of his research career, and an active field of research today.

We now briefly sketch Professor Plans' personality and Mathematical work. The latter will necessarily be incomplete since at this very moment he is vigorously active and producing new results. For example, he is at present giving advice to four students for their doctoral dissertations.

Born in Madrid, he was the sixth of seven children. He was brought up in the cheerful atmosphere that characterizes such large families.

Soon after there came harder times; the social environment of his country was full of tension, the religious persecution started, and the growing insecurity and anxiety ended up in the civil war (1936-39). In fact, the Jesuit school he attended in Madrid was burnt down after taking the children out of their classrooms in May 1931 one month after the proclamation of the Republic. His father died in 1934 and during the war the family lost their house and their material belongings. So, when he started college at the University of Barcelona he had to find part-time jobs to contribute to the family income. Among them, he cooperated in the "Seminario Matemático" conducted by Orts Aracil, and tutored at a private Engineering school designed to complement the curriculum taught at the public university.

He graduated as a Mathematician in the University of Madrid where he had been in his last year of college. In his family there had been already a long tradition of university professors; his grandfather had been a professor at the University of Barcelona and later on he had been among the founders of the school of Pharmacy at the University of Santiago de Compostela.

His father, a talented man who died in his early fifties, held a chair of Physics at the University of Madrid; at the same time he worked intensively in the so-called "Laboratorio Matemático" of the "Junta para Ampliación de Estudios". This was an organization created in order to push forward the impulse given by Rey Pastor to the Mathematical Sciences. He was also an active member of the "Sociedad Matemática Española" and a founder of the "Revista", a journal of those days.

As I prepared these notes, trying to find out precise data of the Plans' family and their environment, there came to my hands a reprint of the homage paid to his father, José María Plans, when he died. Reading it I realized to what extent the same words of praise could have been said of Antonio Plans, and the great resemblance between father and son. After glossing his scientific achievements, his contacts with Levi-Civita, Einstein and many of the personalities of his time, and his influence in the scientific world, Puig Adam underlines heavily his modesty, his great concern for everybody, striving to make others shine, while he himself remained voluntarily in the background. As a summary he says that José María Plans was a wise man, a magister and a holy man, by his christian virtues and his attitude towards life.

Without doubt Antonio Plans has inherited the goodness and the sharp talent of his father, together with the soft manners of his mother. He is a brilliant mathematician and he is also a loyal friend. By his side everyone feels comfortable and self confident because he always finds one's virtues and reasons to be appreciated and trusted. All who happen to come in contact with him (administrative or maintenance staff members, students etc.) are not strangers to him; he knows their names, personal details, their needs..., and he always has an open smile and friendly attitude for everyone. He is also a very organized person, for whom all the small affairs of everyday life are important and deserve consideration.

His professional activity has always been linked with teaching. Since 1957 he has held a chair at the University of Zaragoza. Two of his most outstanding features are his great enthusiasm for, and the attention paid to his students. He has devoted a great deal of his life time to his undergraduate students, for whom his unlimited patience, his concern to make the hard matters easy, his clear and neat exposition in lectures, and the personal communication he established whatsoever the circumstances -overcrowded lessons, for instance- were well known.

This is even more the case with his postgraduate students. He has generously given up his time, encouraging them to go further and beyond in their work. He helps out to unthinkable limits. One of them said, in front of a tribunal of mathematicians judging his examination to become a regular professor "I owe him what I am now". This opinion could be suscribed, I think, by all of us, his students.

We give now a brief account of the ideas underlying or giving rise to his mathematical work, and the list of his publications to date. From now on all citations are referred to the latter. We do not attempt completeness due to a) the many items he has dealt with, and b) the intensive work he is developing at present in quite a number of projects.

It is very remarkable how Geometry has been his leading line; in fact he has an special ability to focus items from a geometrical point of view.

The existence of a regular basis (also called strong M-basis) in every separable Banach space is an open question which has motivated some of his research. A fundamental sequence (x_n) in a Banach space B is a <u>regular basis</u> if there exists a total sequence of functionals $(f_n) \subset B^*$ such that (x_n, f_n) is a biorthogonal system and for every subsequence $(n_k) \subset \mathbb{N}$, $[f_{n_k}]_\perp = [x_j]_{j \in \mathbb{N}-(n_k)}$ ([ ] stands for closed linear hull and $\perp$ for orthogonal).

Regular bases are more general than Schauder bases, and are among M-bases, which are known to exist in every separable Banach space. Papers in this line are: (1983 a), (1985), (1987 b & c), (1988) and the doctoral dissertation advised [8].

An idea of his is the so called "linear simplification" which roughly speaking consists of searching a "good" subsequence out of a given sequence. This "goodness" means amongst other things, that it must have the same closed linear span as the given one, and have some other property such as being minimal, or M-basis, or regular basis, etc. The doctoral dissertations [4] and [12] deal with it, and also (1968 a) , (1969 b), (1976), (1987 b & c).

An auxiliary tool developed by him with one of his students, Andrés Reyes, was the lattice properties of closed linear spans of the subsequences of a given sequence, as is reflected in [8] and (1969 a), (1976), (1981), (1983 a), (1983 b). It seems appropriate here to make an special mention of Andrés Reyes, a gifted student of Plans who died prematurely in a car accident, and whose memory is very beloved to Antonio Plans and to all the members of the Department.

The linear operators acting on a Hilbert Space have been studied by A. Plans from different points of view. In relation with summability he has advised the doctoral dissertations [5] and [9] and the papers (1975 c, d & e) also deal with it. The papers (1977 a & b), (1986 a) are concerned with properties of the images of orthonormal bases. Another item studied by him are those operators A which admit a representation given by $A=\lambda U+C$, with $\lambda \in \mathbb{C}$, U and C unitary and compact operators respectively. He relates them to the hypervolume of a sequence, defined by him, and with Bari systems of vectors and rays. In this line he has advised [3] and [7] and written the papers (1964 a), (1965 b), (1966), (1967 b), (1975 a), (1979 b), (1985 b), (1988 b).

The idea of convergence has also played an important role in his research. Dealing particulary with it are (1959 c), (1964 a), (1967 c), (1973 b) and [6].

Some of his papers are concerned with General Topology, Geometric Algebra and Geometric Topology. In fact a well known result of Knot Theory is the so called Plans theorem, which refers to the homology groups of the branched cyclic coverings of a knot.

At present he is writing lecture notes on what he has called "Espacios de Apoyo". (Shuttle spaces, or Leaning spaces) which have their origin in an old result of Bertini in Projective n-dimensional Geometry. Before this he developed a theory of "unit position", as is reflected in the advising of [8] and [11], and has studied properties of the infinite-dimensional affine spaces as in [13] and (1987 a).

Finally we mention that he is a member of the "Real Academia de Ciencias" of Madrid and of Zaragoza and that he has travelled and lectured in several countries of Europe in many occasions. Presently, he is Professor Emeritus at the University of Zaragoza.

Acknowledgements:

I am indebted to Pedro Plans for a detailed conversation with him, before writing this article, to Jose María Montesinos, Enrique Outerelo, Esteban Induráin and David Tranah, for reading the article and making corrections on it.

References.

Puig Adam, P.: *Jose Maria Plans. Maestro.* Las Ciencias, año II, nº 2.

PUBLICATIONS OF ANTONIO PLANS.

(1946 a) Adriano María Legendre. Matemática Elemental. VI p. 1-3.

(1946 b) Espacio de Hilbert de n dimensiones. Rev. Acad. Ciencias Madrid 40, p. 1-70.

(1947) Espacio de Hilbert. Rev. Acad. Ciencias Madrid 41, p. 197-257.

(1948) Operadores lineales en el espacio de Hilbert y su espectro. Rev. Acad. Ciencias Madrid 42, p. 309-391.

(1949) with J. Teixidor. Conferencias sobre los operadores lineales en el espacio de Hilbert. Obras completas de Gaston Julia. IV. p. 479-530.

(1952 a) Sobre la aproximación dimensional en el espacio de Kuratowski, Rev. Acad. Ciencias Madrid, XLVI, p. 303-306.

(1952 b) Algunas propiedades lineales de las matrices acotadas. Rev. Acad. Ciencias Madrid XLVI, p. 273-302.

(1952 c) Ensayo de un álgebra lineal infinita en el campo de las matrices acotadas. Coll. Math. V, p. 3-47.

(1952 d) Una forma algebraica de la dimensión de Urysohn en el espacio de Kuratowski. Rev. Acad. Ciencias Madrid VII, p. 47-50.

(1952 e) Sobre los invariantes métrico-afines de las formas cuadráticas. Gaceta Matemática IV, 7 y 8 p. 248-253.

(1953) Aportación al estudio de los grupos de homología de los recubrimientos cíclicos ramificados correspondientes a un nudo. Rev. Acad. Ciencias Madrid XLVII, p. 161-193.

(1956 a) Estudio sintético del espacio proyectivo de base no finita numerable. Publicaciones C.S.I.C, p. 1-109.

(1956 b) Primeras propiedades de las hipercuádricas en el espacio proyectivo con base no finita numerable. Rev. Mat. Hisp.-Amer. 16, p. 11-27.

(1957 a) Aportación a la homotopía de sistemas de nudos. Rev. Mat. Hisp.-Amer. 17, p. 1-14.

(1957 b) Un sistema de axiomas para el anillo de las matrices infinitas acotadas reales. Coll. Math. 9, p. 35-40.

(1957 c) Una estructura reticular del anillo de las matrices infinitas acotadas reales. Coll. Math. 9, p. 87-104.

(1958) El Tiempo. Actas de la Primera Reunión de Aproximación Filosófico-Científica, p. 133-150.

(1959 a) Propiedades angulares de los sitemas heterogonales. Rev. Acad. Ciencias Zaragoza 14, p. 5-18.

(1959 b) El Espacio. Actas de la Segunda Reunión de Aproximación Filosófico-Científica, p. 132-138.

(1959 c) Zerlegung von Folgen im Hilbertraum in Heterogonalsysteme. Arch. Math. 10, p. 304-306.

(1961 a) Resultados acerca de una generalización de la semejanza en el espacio de Hilbert. Coll. Math. 13, p. 241-258.

(1961 b) Propiedades angulares de la convergencia en el espacio de Hilbert. Rev. Mat. Hisp.-Amer. 21, p. 100-109.

(1962) Sobre los operadores lineales acotados en relación con la convergencia de variedades lineales. Coll. Math. 14, p. 269-274.

(1963 a) Los operadores acotados en relación con los sistemas asintóticamente ortogonales. Coll. Math. 14, p. 269-274.

(1963 b) Momento actual de la investigación matemática. Cuadernos de Ciencias. Universidad de Zarazoga I, 6. p. 3-11.

(1964 a) Sobre la convergencia débil en el espacio de Hilbert. Rev. Acad. Ciencias Zaragoza 19, p. 69-73.

(1964 c) Sobre una caracterización de los operadores acotados y extensiones suyas, mediante sistemas heterogonales de rayos. Actas V. R.A.M.E. p. 99-102.

(1965 a) Una caracterización de los operadores biunívocos de doble norma finita. Actas de la IV R.A.M.E. p. 119-122.

(1965 b) Sobre un determinante infinito definido mediante un operador de doble norma finita. Actas de la IV R.A.M.E. p. 123-129.

(1966) Sobre la representación de un operador lineal acotado mediante operadores isométricos y completamente continuos. Rev. Mat. Hisp.-Amer. 26, p. 202-206.

(1967 a) Sobre la continuidad uniforme angular de los operadores lineales acotados biunívocos. Actas VI R.A.M.E. p. 76-80.

(1967 b) Nuevos resultados sobre una generalización de la semejanza en el espacio de Hilbert. Actas de la VI R.A.M.E. p. 81-83.

(1967 c) Sobre el conjunto de los rayos del espacio de Hilbert. Cuadernos de Ciencias, Universidad de Zaragoza 19, p. 5-11.

(1968 a) Simplificación lineal en el espacio de Hilbert. Rev. Mat. Hisp.-Amer. 28 p. 196-199.

(1968 b) Notas sobre la compacidad de los rayos del espacio de Hilbert. Actas de la IX R.A.M.E. p. 166-170.

(1969 a) Dependencias lineales en el espacio de Hilbert. Pub. Sem. Mat. García de Galdeano 10, p. 153-161.

(1969 b) Extensión de la dependencia lineal en el espacio de Hilbert. Actas de la VIII R.A.M.E. p. 90-92.

(1969 c) Espacio de Hilbert. Rev. Acad. Ciencias de Zaragoza. p. 1-40.

(1971) La topología débil del conjunto de los rayos del espacio de Hilbert, definida mediante hiperplanos. Rev. Acad. Ciencias Zaragoza 26, 3 p. 525-528.

(1972) Definición de una topología en el conjunto de los rayos del espacio de Hilbert. Actas X R.A.M.E. p. 123-129.

(1973 a) Una base de la topología débil en el conjunto de rayos del espacios de Hilbert, definida mediante hiperplanos. Actas de la XI R.A.M.E. p. 146-154.

(1973 b) Sobre los sistemas p-heterogonales. Actas I Jornadas Mat. Hisp.-Lusas. p. 270.

(1975 a) Caracterización de los operadores lineales biunívocos en el espacio de Hilbert, que intercambian sistemas completos de rayos de Bari. Rev. Mat. Hisp.-Amer. 5, p. 158-166.

(1975 b) A generalization of heterogonal systems. Arch. Math. 26, 4, p. 398-401.

(1975 c) Sobre la sumabilidad en norma en bases débiles en el espacio de Hilbert, mediante operadores acotados. Actas I Sem Math. Franco-Espagnole.

(1975 d) Transformación de sistemas heterogonales completos de rayos mediante operadores de Hilbert-Schmidt y nucleares. Actas I Sem. Math. Franco-Espagnole.

(1975 e) with Burillo, P. Operadores nucleares y sumabilidad en bases ortonormales. Actas I Sem Math. Franco-Espagnole.

(1976) Sistemas de vectores con núcleo nulo en el espacio de Hilbert. Actas de la XII R.A.M.E. p. 145-150.

(1977 a) Imagen de un sistema ortogonal completo de rayos por un operador lineal acotado. Coll. Math. 28, 3, p. 177-183.

(1977 b) Una caracterización de los operadores completamente continuos mediante sistemas ortonormales. Actas IV Jornadas Mat. Hisp.-Lusas p. 359-363.

(1979 a) Heterogonal systems of subspaces. Proc. of the IV Int. Colloq. of Diff. Geometry, p. 225-233.

(1979 b) Sobre los sistemas ξ-heterogonales. Rev. Univ. Santander 2, p. 607-646.

(1980 a) Comportamiento de los operadores acotados en los sistemas ortogonales. Pub. Mat. Univ. Aut. Barcelona 21, p. 217.

(1980 b) Espacios de apoyo. Pub. Mat. Univ. Aut. Barcelona, 21. p. 127-128.

(1981) Propiedades inducidas por subsistemas de complemento infinito en el espacio de Hilbert. Actas VIII Jornadas Mat. Luso-Esp. p. 353-355.

(1983 a) with Reyes, A., On the geometry of sequences in Banach spaces. Arch. Math. 40 p. 452-458.

(1983 b) Another proof of a result of P. Terenzi. Proc. II Int. Conf. on Operator Algebras p. 200-202.

(1985 a) A characterization of basic sequences in Banach spaces. Symp. Aspects of Positivity in Funct. Anal. Tübingen, p. 115-116.

(1985 b) Sobre semejanza asintótica de operadores en el espacio de Hilbert. Act. Reunión de Probabilidad y Espacios de Banach, Zaragoza.

(1985 c) On the geometry of sequences in Banach spaces. Rendiconti Sem. Mat. e Fis. di Milano, LV.

(1986 a) with Martín-Peinador, E. Sistemas L de rayos y sumabilidad. Hom. Prof. Botella. Univ. Complut. Madrid. p. 203-218.

(1986 b) with García-Castellón, F. and Induráin, E. Propiedades universales de sucesiones en el espacio de Banach. Mem. Acad. Ciencias Madrid, 31, p. 1-25.

(1987 a) Un problema geométrico y afín de cuasi-complementariedad en espacios de Banach. Hom. Prof. Cid. Univ. Zaragoza, p. 173-180.

(1987 b) with García-Castellón, F. On linear simplification in Banach spaces I. Boll. U.M.I. (7) 1A. p. 87-95.

(1987 c) with García-Castellón, F. On linear simplification in Banach spaces II. Boll. U.M.I. 1A p. 367-373.

(1988 a) with Induráin, E. and Reyes, A. Notas sobre Geometría de Sucesiones en Espacios de Banach. Univ. Extremadura-Zaragoza. p. 1-262 (Book).

(1988 b) with Martín-Peinador, E. ; Induráin, E. and Rodés, A. Two geometric constants for operators acting on a separable Banach space. Rev. Mat. Univ. Complutense 1; 1,2,3, p. 25-32.

(1987) with García-Castellón and Induráin, E. Bases a través de sistemas minimales uniformes. Pub. Sem. Mat. G. Galdeano. Univ. Zaragoza 1-5.

(1988) with Cuadra, J.L.; García-Castellón, F.; and Induráin, E. Some properties of sequences with 0-kernel in Banach spaces. (To appear in Rev. Roum. Sci.)

DOCTORAL DISSERTATIONS ADVISED.

[1] (1965) "Propiedades de una función de valores enteros, definida en el conjunto de las multiálgebras finitas". by Bartolomé Frontera.

[2] (1970) "Sobre el conjunto de los rayos del espacio de Hilbert". by Victor Onieva.

[3] (1974) "Hipervolúmenes, su límite y propiedades de permanencia en los operadores lineales acotados del espacio de Hilbert" by Pedro Burillo.

[4] (1975) "Sobre sucesiones en los espacios de Hilbert y Banach. Problema de la simplificación lineal topológica" by Jose Luis Cuadra.

[5] (1977) "La sumabilidad absoluta en los operadores lineales acotados del espacio de Hilbert" by Elena Martín-Peinador.

[6] (1978) "Convergencias en el conjunto de los subespacios del espacio de Hilbert" by M. Carmen de las Obras-Loscertales.

[7] (1978) "Hipervolúmenes, su convergencia y sumabilidad en los operadores lineales acotados del espacio de Hilbert" by Eusebio Corbacho.

[8] (1980) "Aspectos reticulares y geométricos de sistemas de vectores en espacios de Banach y de Hilbert" by Andrés Reyes.

[9] (1980) "Ideales de operadores lineales acotados y sumabilidad en el espacio de Hilbert" by Alvaro Rodés.

[10] (1983) "Involuciones en espacios topológicos" by Paulino Ruiz de Clavijo.

[11] (1985) "Sobre geometría de sucesiones en espacios de Banach. Posición unidad" by Esteban Induráin.

[12] (1985) "Geometría de los hiperplanos asociados a una sucesión en un espacio de Banach. Aplicación al problema de la simplificación lineal" by Felicísimo García-Castellón.

[13] (1985) "Sucesiones en la geometría afín del espacio de Banach" by María Jesús Chasco.

INFINITE DIMENSIONAL GEOMETRIC MODULI AND TYPE-COTYPE THEORY

V. Milman and A. Perelson
The Raymond and Beerley Sackler
Faculty of Exact Sciences
Tel Aviv University
Tel Aviv, Israel

1. INTRODUCTION

In this paper, we study an infinite dimensional normed space B by combining the geometric approach, in the spirit of [M1], with type-cotype theory, in the spirit of Maurey-Pisier [MaPi]. Of course, all linear topological properties of B are defined by the structure of its unit sphere $S(B) = \{x \in B: \|x\| = 1\}$. However, the infinite dimensional unit sphere is a difficult to visualize object. Therefore, it is natural to construct simply computable and visual numerical characteristics of the sphere (often called moduli) which do not define our space uniquely but describe some topological properties of the space. The first such modulus was the modulus of convexity, introduced in 1936 by Clarkson [Cl],

$$\delta_B(\varepsilon) = \inf\left\{\left(1 - \frac{\|x+y\|}{2}\right) \,\middle|\, x,y \in S(B),\ \|x-y\| = \varepsilon\right\},\quad 0 \le \varepsilon \le 2.$$

Not long after, D.P. Milman [Mil] showed that a uniformly convex space B, i.e., B such that $\delta_B(\varepsilon) > 0$ for $\varepsilon > 0$, is reflexive. V.L. Šmulian [Sm] and M.M. Day [D] have proved that the property of uniform convexity is dual to the property of uniform smoothness. The modulus of smoothness of space B is the function (for $\varepsilon>0$)

$$\rho_B(\varepsilon) = \frac{1}{2}\sup\left\{\|x+y\| + \|x-y\| - 2 \,\middle|\, \|x\| = 1,\ \|y\| = \varepsilon\right\}.$$

We call the space B uniformly smooth if $\rho_B(\varepsilon)=o(\varepsilon)$. J. Lindenstrauss [Li] has pointed out the exact duality relations between $\delta_B(\varepsilon)$ and $\rho_B(\eta)$. Also, some properties of unconditionally convergent series were connected with these moduli $\delta_B(\varepsilon)$ and $\rho_B(\eta)$ by M. I. Kadeč [K] and J. Lindenstrauss [Li].

This short list of results essentially gives all the available information, which was obtained using both moduli. (Also, the moduli by themselves and especially their local variants became an important tool in the study of normed spaces.) The limiting ability of these moduli in the linear topological study of a normed space is, of course, connected with the fact that they are constructed by a family of 2-dimensional subspaces of B. However, the only space defined completely by this family of subspaces is a Hilbert space.

To avoid such an insufficiency of 2-dimensional moduli, the first author developed, at the end of the '60's, the language of β- and δ-moduli, reflecting the local geometric structure of the unit sphere S in an infinite dimensional sense. Let $\mathcal{B}$ be some family of subspaces of the space B. For any $x\in S(B)$, we define local moduli, β-modulus (the lower one) and δ-modulus (the upper one) of the family $\mathcal{B}$ by

$$(1.1) \qquad \beta(\varepsilon;x,\mathcal{B})=\sup_{E\in\mathcal{B}}\ \inf_{y\in S(E)}\left(\|x+\varepsilon y\|-1\right);\ \delta\left(\varepsilon;x,\mathcal{B}\right)=\inf_{E\in\mathcal{B}}\ \sup_{y\in S(E)}\left(\|x+\varepsilon y\|-1\right),$$

where $S(E)=\{x\in E \mid \|x\|=1\}$ is the unit sphere of the space E. These moduli reflect two averaging procedures. Let $f(x)$, $x\in S(B)$, be a bounded real-valued function on the sphere. Then we define β- and δ-averaging with respect to the family $\mathcal{B}$ by

$$\beta[f,\mathcal{B}]=\sup_{E\in\mathcal{B}}\ \inf_{x\in S(E)} f(x) \quad\text{and}\quad \delta[f;\mathcal{B}]=\inf_{E\in\mathcal{B}}\ \sup_{x\in S(E)} f(x).$$

We only apply these averagings to the family of functions $\|x+\varepsilon y\|-1$ of two variables x and y, $\|x\|=\|y\|=1$, and a parameter $\varepsilon>0$. (However, we use different families $\mathcal{B}$). The first time we average by variable y and obtain the local moduli (1.1). Then we average them by the remaining variable x and we receive the global (uniform) moduli

$$\beta[\beta(\varepsilon;x,\mathcal{B});\mathcal{B}]=\beta\beta(\varepsilon;\mathcal{B}),\quad \delta[\beta(\varepsilon;x,\mathcal{B});\mathcal{B}]=\delta\beta(\varepsilon;\mathcal{B})$$

and

$$\beta[\delta(\varepsilon;x,\mathcal{B});\mathcal{B}]=\beta\delta(\varepsilon;\mathcal{B}),\quad \delta[\delta(\varepsilon;x,\mathcal{B});\mathcal{B}]=\delta\delta(\varepsilon;\mathcal{B}).$$

Note that, e.g., if $\mathcal{B}$ is the family $\mathcal{E}_1$ of all 1-dimensional subspaces of B, then $\delta\delta(\varepsilon;\mathcal{E}_1)$ is equivalent (for $\varepsilon\to 0$) to the modulus of convexity $\delta_B(\varepsilon)$ and, similarly, $\beta\beta(\varepsilon;\mathcal{E}_1)$ is equivalent to the modulus of smoothness $\rho_B(\varepsilon)$.

In this paper we will deal with three families. The first one is the family $\mathcal{B}^0$ of all subspaces of a finite codimension. In this case, we often prefer to emphasize the space B and write $\beta^0\beta^0(\varepsilon,B)$ instead of $\beta\beta(\varepsilon;\mathcal{B}^0)$ (and similarly for other moduli). However, we omit a family $\mathcal{B}$ and simply put $\beta\beta(\varepsilon)$ if the family of subspaces we are using is clear from the context. Also note that $\mathcal{B}^0(E)$ denotes the family of all subspaces of the space E of finite codimension in E.

A few examples:

i. $\beta^0(\varepsilon;x;\ell_p) = \delta^0(\varepsilon;x;\ell_p) = \sqrt[p]{1+\varepsilon^p} - 1 \simeq \varepsilon^p/p$ (for any $x\in S(\ell_p)$) and $\beta^0(\varepsilon;x;c_0)=\delta^0(\varepsilon;x;c_0)=\max(0,\varepsilon-1)$.

ii. $\beta^0(\varepsilon;x;C[0,1]) = \max(0,\varepsilon-2)$, $\delta^0(\varepsilon;x;C[0,1])=\varepsilon$.

iii. $\beta^0(\varepsilon;x;L_1[0,1])=0$ $(0\le\varepsilon\le 1)$, $\delta^0(\varepsilon;x;C[0,1])=\varepsilon$.

iv. There exist constants $C_1(p)$ and $C_2(p)>0$ such that, for $1 < p \le 2$ and $0 \le \varepsilon \le 1$ and any $x\in S(L_p)$

$$C_2(p)\varepsilon^2 \le \beta^0(\varepsilon;x;\ L_p[0,1]) \le C_1(p)\varepsilon^2,\quad C_2(p)\varepsilon^p \le \delta^0(\varepsilon;x;L_p) \le \le C_1(p)\varepsilon^p;$$

for $2 \le p < \infty$ and $0 \le \varepsilon \le 1$

$$C_2(p)\varepsilon^p \le \beta^0(\varepsilon;x;\ L_p[0,1]) \le C_1(p)\varepsilon^p \quad\text{and}$$

$$C_2(p)\varepsilon^2 \le \delta^0(\varepsilon;x;\ L_p) \le C_1(p)\varepsilon^2.$$

The second family of subspaces which we use in this paper in section 4 is produced by a minimal system (or a basic sequence) $X = \{x_i\}_1^\infty \subset B$. Then $\mathcal{B}=\mathcal{B}(X)$ is the family $\left\{\overline{\text{span}}\{x_i\}_{i=n}^\infty\right\}_{n=1}^\infty$.

The first two families satisfy the so called filtration condition: $\forall E_i \in \mathcal{B}$, $i=1,2$ $\exists E_3 \subset (E_1 \cap E_2)$ and $E_3 \in \mathcal{B}$. This property plays an important role and also simplifies many problems. The third family and subspaces which we use in section 5 do not satisfy the filtration property and therefore they have to be constructed in a slightly different manner. This is the family $\mathcal{B}_0$ of all subspaces of B of finite (non-trivial) dimension. We also use the natural partition of $\mathcal{B}_0 = \bigcup_{K\geq 1} \mathcal{B}_k$ where $\mathcal{B}_k$ is the family of all k-dimensional subspaces of B. Define finite dimensional moduli

$$\beta_k(\varepsilon;x) = \beta(\varepsilon;x,\mathcal{B}_k) \quad \text{and} \quad \delta_k(\varepsilon;x) = \delta(\varepsilon;x,\mathcal{B}_k),$$

and

$$\beta_0(\varepsilon;x) = \lim_{k\to\infty} \beta_k(\varepsilon;x), \quad \delta_0(\varepsilon,x) = \lim_{k\to\infty} \delta_k(\varepsilon;x).$$

Similarly, we construct the global moduli $\beta_0\beta_0(\varepsilon) = \lim_{n\to\infty} \beta[\beta_0(\varepsilon,x);\ \mathcal{B}_n]$ and $\delta_0\delta_0(\varepsilon)$.

Between 1967 and 1971, the first author pursued an investigation (see survey [M1]) to show that a local geometric structure of S influences (and sometimes defines) a global topological property of a normed space. It was shown that the language of β- and δ-moduli, introduced above, connects local geometry of the sphere of the space with different topological properties of the space, such as reflexivity, isomorphism to some dual space, a property of a separable space having a separable dual, etc. It helps to compare different topologies and, for example, gives the complete and simple condition when the weak and the strong topologies on the sphere coincide.

Let us recall one example of such a result.

<u>Theorem</u> ([M1]). *Let* B *be a separable Banach space.*

a) *If* $\beta^0(\varepsilon_0;x,B) = 0$ *for some* $\varepsilon_0 > 0$ *and any* $x \in S(B)$, *then* B *is not isomorphic to a dual space, and moreover,* B *is not isomorphic to a subspace of a separable dual space.*

b) *If* $\delta^0(\varepsilon; x, B) = \varepsilon$ *then* B^* *is not separable.*

Another direction of geometric investigation of normed spaces was succesfully developed during the last 10-15 years, the so-called Local Theory. Its purpose is a study of the structure of finite dimensional subspaces of a given space, quotient spaces, projections and so on (see, e.g., [MSch]).

In this paper we want to revive the old local geometric approach from [M1] by connecting it with new methods of Local Theory. We show below that local geometric behaviour of the sphere S implies some knowledge of the structure of finite dimensional subspaces, i.e., the kind of global properties in the geometric sense which we call the local one in the sense of the linear structure.

We use below the notations and results from [M1] on geometry of the unit sphere and some properties of sequences in a normed space; we also refer to [MSch] on the results on Local Theory, especially on elements of a type-cotype theory of normed spaces. We usually write $a \overset{\Theta}{\simeq} b$ instead of $|a-b| \le \Theta$.

2. DESCRIPTION OF THE MAIN RESULTS AND DISCUSSION

We start with the following results which will be proved in section 3.

Theorem 2.1. *Let* B *be an infinite dimensional Banach space and* $\mathcal{B}^0$ *the family of all closed subspaces of* B *of finite codimension.*

i) *If there exists* p, $1 \le p \le 2$, *such that*

$$\beta\beta(\varepsilon, \mathcal{B}^0) \ge c\varepsilon^p$$

for $0 < \varepsilon < 1$ *and some constant* c, *then* ℓ_p *is finitely represented in* B *(i.e., for any* $\delta > 0$ *and an integer n,* B *contains an* $(1+\varepsilon)$*-isomorphic copy of* ℓ_p^n*; see* [MSch], *Chapter 11, for the exact definitions).*

ii) *If there exists* $q \geq 2$ *such that*

$$\delta\delta(\varepsilon, \mathcal{B}^0) \leq c\varepsilon^q$$

for $0 < \varepsilon < 1$ *and some constant* c, *then there exists* $q_0 \geq q$ *and* ℓ_{q_0} *is finitely represented in* B.

iii) *If there exists* $\varepsilon_0 > 0$ *such that* $\delta\delta(\varepsilon_0, \mathcal{B}^0) = 0$ *then* ℓ_∞ *is finitely represented in* B.

The two important partial cases of this theorem are, indeed, complementary to some results from [M1]: the case p=1 in part i) of the theorem and part iii) of this theorem (which is, of course, a partial case of Theorem 2.1.ii for $q=\infty$). We referred to the following facts.

Theorem 2.A ([M1], Theorem 5.5.a and b). **a.** *If* $\beta\delta(1;\mathcal{B}^0) = 0$ *then* B *contains an isomorphic copy of* c_0; *if* $\delta\beta(\varepsilon;\mathcal{B}^0) = \varepsilon$ *for* $0 \leq \varepsilon \leq 1$; *then* B *contains an isomorphic copy of* ℓ_1.

b. *If* B *contains an isomorphic copy of* c_0 *then* $\beta\beta(\varepsilon, \mathcal{B}^0) = 0$ *for* $0 \leq \varepsilon \leq 1$; *if* B *contains an isomorphic copy of* ℓ_1 *then* $\delta\delta(\varepsilon, \mathcal{B}^0) = \varepsilon$ *for* $0 \leq \varepsilon \leq 1$.

The next results which we choose to discuss in this section will be proved in section 4.

Theorem 2.2. *Let* $X = \{x_i\}_1^\infty$ *be a minimal system of* B *with the total dual* $X^* = \{x_1^*\}_1^\infty \subset B^*$. *If there exists a constant* C *such that for* $0 \leq \varepsilon \leq 1$

$$\delta\delta(\varepsilon, \mathcal{B}(X)) \leq C\beta\beta(\varepsilon, \mathcal{B}(X))$$

then either there exists $\varepsilon_0 > 0$ *and* $\beta\beta(\varepsilon_0, \mathcal{B}(X)) = 0$ *and in this case* ℓ_∞ *is finitely representable in the space* B *by blocks of the sequence* X, *or there exists* p, $1 \leq p < \infty$, *such that* $\beta\beta(\varepsilon;\mathcal{B}(X)) = O(\varepsilon^p)$ *and* ℓ_p *is finitely representable in the space* B *by blocks of the sequence* X.

This theorem should be compared with two known results.

Theorem ([M1]), Theorem 4.2 and Corollary 4.5). **i)** *Let there exist constants* $C_2 \geq C_1 > 0$, $\varepsilon_0 > 0$ *and a function* $\psi(\varepsilon)$ *such that for any* $x \in S(B)$

$$C_1\psi(\varepsilon) \leq \beta(\varepsilon;x;\mathcal{B}^0) \leq \delta(\varepsilon;x;\mathcal{B}^0) \leq C_2\psi(\varepsilon).$$

Then either there exists $\varepsilon_1 > 0$ *and* $\psi(\varepsilon) = 0$ *for* $0 \leq \varepsilon \leq \varepsilon_1$ *and B contains an isomorphic copy of* c_0, *or there exists* $p \geq 1$ *and constants* $A_2 \geq A_1 > 0$ *such that*

$$A_1\varepsilon^p \leq \psi(\varepsilon) \leq A_2\varepsilon^p$$

for $0 \leq \varepsilon \leq \varepsilon_0$ *and B contains an isomorphic copy of* ℓ_p.

ii) *Let* $\beta\beta(\varepsilon;\mathcal{B}^0) = \delta\delta(\varepsilon;\mathcal{B}^0) = \psi(\varepsilon)$. *Then either* $\psi(\varepsilon) = 0$ *for* $0 \leq \varepsilon \leq 1$ *and B contains an* $(1+\delta)$*-isomorphic copy of* c_0 *for any* $\delta > 0$, *or there exists* $p \geq 1$ *and* $\psi(\varepsilon) = (1+\varepsilon^p)^{1/p}-1$ *and B contains an* $(1+\delta)$*-isomorphic copy of* ℓ_p *for any* $\delta > 0$.

Also one has to note in this connection the properties of an infinite dimensional sphere reflected in the following proposition.

Proposition ([M1], Theorem 5.9). **i)** *If* $\delta\beta(\varepsilon_0;\mathcal{B}^0) = 0$ *for some* $\varepsilon_0 > 0$ *then* $\beta\beta(\varepsilon:\mathcal{B}^0) = 0$ *for* $0 \leq \varepsilon \leq 1$.

ii) *If* $\lim_{\varepsilon\to 0} \frac{\beta\delta(\varepsilon;\mathcal{B}^0)}{\varepsilon} = c > 0$ (*this* lim *always exists*) *then* $\delta\delta(\varepsilon:\mathcal{B}^0) = \varepsilon$ *for* $\varepsilon > 0$.

In section 5 we improve Theorem 2.1 using the finite dimensional moduli $\beta_0\beta_0(\varepsilon)$ and $\delta_0\delta_0(\varepsilon)$. It is known ([M1], Proposition 3.2 and [M3]) that for any infinite dimensional Banach space B

$$\delta\delta(\varepsilon,B^o) \geq \beta_0\beta_0(\varepsilon) \geq \delta_0\delta_0(\varepsilon) \geq \beta\beta(\varepsilon,B^0).$$

Therefore the following statement strengthens Theorem 2.1.

Theorem 2.3. **i)** *If there exist constants* $c < \infty$, $\varepsilon_0 > 0$ *such that for some* $1 \le p \le 2$ *and* $0 < \varepsilon \le \varepsilon_0$

$$\delta_0\delta_0(\varepsilon) \ge c\varepsilon^p$$

then ℓ_p *is finitely representable in* B.

ii) *If there exist* $C < \infty$, $\varepsilon_0 > 0$ *such that for some* $q \ge 2$ *and* $0 < \varepsilon \le \varepsilon_0$

$$\beta_0\beta_0(\varepsilon) \le c\varepsilon^q$$

then ℓ_{q_0} *is finitely representable in* B *for some* $q_0 \ge q$.

iii) *If* $\beta_0\beta_0(\varepsilon_0) = 0$ *for some* $\varepsilon_0 > 0$, *then* ℓ_∞ *is finitely representable in* B.

3. ASYMPTOTIC BEHAVIOR OF GLOBAL β- AND δ-MODULI

The next theorem gives a connection between the global moduli of a Banach space B, from one side and its type and cotype from the other side. In this section we use only the family of subspaces $\mathcal{B}^0$ and so we write $\beta\beta(\varepsilon)$ instead of $\beta\beta(\varepsilon;\mathcal{B}^0)$ and similarly $\delta\delta(\varepsilon)$, $\beta(\varepsilon;x)$ and so on.

Theorem 3.1. *Let* B *be an infinite dimensional Banach space of type* $p \ge 1$ *with the type* p *constant* T_p *and cotype* $q < \infty$ *with the cotype* q *constant* C_q. *Then, for a universal constant* C,

$$\overline{\lim_{\eta\to 0}} \frac{\beta\beta(\eta)}{\eta^p} \le C\cdot T_p^p \tag{3.1}$$

and

$$\underline{\lim_{\eta\to 0}} \frac{\delta\delta(\eta)}{\eta^q} \ge \frac{1}{q}\left(\frac{1}{3C_q}\right)^q. \tag{3.2}$$

In the proof of Theorem 3.1 we use the standard technique from [M1]. The following lemma is quite obvious.

Lemma 3.2. *For every* $\Theta'>0$ *there exists a subspace* $E_{\Theta'} \in \mathcal{B}^0$ *such that for every* $\varepsilon > 0$, $0 \le \varepsilon \le 1$, *we have*

$$(3.3) \qquad \beta\beta(\varepsilon) \overset{\Theta'}{\simeq} \inf_{x \in S(E_{\Theta'})} \beta(\varepsilon; x),$$

and

$$(3.4) \qquad \delta\delta(\varepsilon) \overset{\Theta'}{\simeq} \sup_{x \in S(E_{\Theta'})} \delta(\varepsilon; x).$$

We also use the following result.

Theorem 3.3. (see [M1], Theorem 4.1). *For every* $\Theta > 0$ *there is a basic sequence* $X = \{x_j\}_1^\infty \subset S(B)$, *such that for every finite scalar sequence* $\{a_j\}_1^k$ *with* $a_1 = 1$ *we have*

$$(3.5) \qquad (1-\Theta)\prod_{j=2}^{k}\left[1+\beta\left(\frac{|a_j|}{\|\sum_{n=1}^{j-1} a_n x_n\|}; \sum_{n=1}^{j-1} a_n x_n\right)\right] \le \left\|\sum_{n=1}^{k} a_n x_n\right\| \le$$

$$\le (1+\Theta)\prod_{j=2}^{k}\left[1+\delta\left(\frac{|a_j|}{\|\sum_{n=1}^{j-1} a_n x_n\|}; \sum_{n=1}^{j-1} a_n x_n\right)\right].$$

Moreover, inequality (3.5) also holds for every basic block sequence $Y \subset S(B)$ *with respect to* X.

Proof of Theorem 3.1. For $\Theta' > 0$ let $E_{\Theta'} \in \mathcal{B}^0$ be a finite codimensional subspace for which (3.3) and (3.4) hold. It is known (see Theorem 3.1.b from [M1]) that for every $x \in E_{\Theta'}$ and $\varepsilon > 0$ we have

$$\beta^0(\varepsilon; x) = \beta(\varepsilon; x; \mathcal{B}^0(E_{\Theta'}))$$

and

$$\delta^0(\varepsilon;x) = \delta(\varepsilon;x;\mathcal{B}^0(E_{\Theta'})).$$

By Theorem 3.3, which we use for the Banach space $E_{\Theta'}$, there is a basic sequence $X \subset E_{\Theta'}$ for which (3.3) (3.4) and (3.5) hold. Let k be an arbitrary natural number. It is well known (see, e.g., [MSch], Chapter 11) that there exists a block basic finite set $\{y_j\}_1^{k+1}$ with respect to the sequence X such that its unconditional constant is at most 3 and, by Theorem 3.3, (3.5) holds. Hence, for every finite sequence of scalars $\{a_j\}_1^{k+1}$ we have

$$\| \sum_{j=1}^{k+1} a_j y_j \| \le 3 \int_0^1 \| \sum_{j=1}^{k+1} r_j(t) a_j y_j \| \, dt, \tag{3.6}$$

where $r_j(t)$, $j = 1,2,\dots$, is the sequence of Rademacher functions on the interval $[0,1]$.

Since B is of type p then, by (3.6), we obtain

$$\| \sum_{j=1}^{k+1} a_j y_j \| \le 3T_p \left(\sum_{j=1}^{k+1} |a_j|^p \right)^{1/p}. \tag{3.7}$$

For $\eta>0$ let us define, by induction, a finite sequence of scalars $\{a_j\}_1^{k+1}$ in the following way, $a_1=1$, and for $j \ge 2$, put

$$\frac{a_j^p}{(3T_p\eta)^p} = \sum_{i=1}^{j-1} a_i^p.$$

Using (3.7) we obtain

$$\frac{a_j}{\|\sum_{i=1}^{j-1} a_i y_i\|} \ge \eta \tag{3.8}$$

It is easily verified by induction, that for $j \ge 2$ we have

$$(3.9) \qquad a_j^p = (1+(3T_p\eta)^p)^{j-2}(3T_p\eta)^p.$$

Using the monotonicity of $\beta(\varepsilon;x)$ (as a function of ε), together with (3.8) we have for $j\geq 2$

$$(3.10) \qquad \beta\left(\eta;\ \sum_{i=1}^{j-1} a_i y_i\right) \leq \beta\left(\frac{a_j}{\left\|\sum_{i=1}^{j-1} a_i x_i\right\|}\ ;\ \sum_{i=1}^{j-1} a_i y_i\right).$$

From (3.3) and (3.10) it follows that for $j \geq 2$, we have

$$1+\beta\left(\eta;\ \sum_{i=1}^{j-1} a_i y_i\right) \geq 1+\inf_{x\in S(E_{\Theta'})} \beta(\eta;x) \geq$$

$$\geq 1+\beta\beta(\eta)-\Theta' \geq (1-\Theta')(1+\beta\beta(\eta)).$$

Suppose that we select Θ' in such a way that $(1-\Theta')^k \geq 1/2$. Using the inequality on the left hand side of (3.5) together with (3.9) we obtain

$$(3.11) \qquad \prod_{j=2}^{k+1}(1+\beta\beta(\eta)) \leq \frac{2}{1-\Theta}\left\|\sum_{j=1}^{k+1} a_j y_j\right\| \leq \frac{2}{1-\Theta}\,3T_p\left(\sum_{j=1}^{k+1} a_j^p\right)^{1/p} =$$

$$= \frac{2}{1-\Theta}\,3T_p\,\frac{a_{k+2}}{3T_p\eta} = \frac{2}{1-\Theta}\,3T_p(1+(3T_p\eta)^p)^{k/p}.$$

Taking the k-th root from both sides of (3.11) we have

$$(3.12) \qquad 1+\beta\beta(\eta) \leq \left(\frac{2}{1-\Theta}\,3T_p\right)^{1/k}(1+(3T_p\eta)^p)^{1/p} \leq$$

$$\leq \left(\frac{2}{1-\Theta}\,3T_p\right)^{1/k}(1+(3T_p\eta)^p/p).$$

The last inequality holds for every natural number k. Taking the limit on both sides of (3.12) we obtain as $k\to\infty$,

$$1+\beta\beta(\eta) \le 1+(3T_p\eta)^p/p.$$

Thus we have

$$\overline{\lim_{\eta\to 0}} \frac{\beta\beta(\eta)}{\eta^p} \le \frac{(3T_p)^p}{p}.$$

This proves the first part of Theorem 3.1. In a similar way we prove the second part of this theorem.

As we did in the proof of the first part, let us assume that $\{y_j\}_1^{k+1}$ is a finite block-basic set with respect to X which has an unconditional constant at most 3. Hence, for every finite sequence of scalars $\{a_j\}_1^{k+1}$, we obtain

$$\| \sum_{j=1}^{k+1} a_j y_j \| \ge 1/3 \int_0^1 \| \sum_{j=1}^{k+1} r_j(t) a_j y_j \| dt \ge \frac{1}{3C_q} \left(\sum_{j=1}^{k+1} |a_j|^q \right)^{1/q}. \tag{3.14}$$

Here C_q is the q-cotype constant of B. As in the first part of the proof, for $\eta > 0$, define, by induction, a sequence of scalars in the following way: $a_1 = 1$ and, for $j \ge 2$, let a_j be defined by the equalities

$$\frac{a_j^q}{\left[\frac{1}{3C_q}\eta\right]^q} = \sum_{j=1}^{j-1} a_i^q. \tag{3.15}$$

It follows from (3.15) that for $j \ge 2$

$$a_j^q = \left(1+\left[\frac{1}{3C_q}\eta\right]^q\right)^{j-2} \left[\frac{1}{3C_q}\eta\right]^q.$$

Let Θ' be selected in such a way that $(1+\Theta')^k \le 2$. As in the first part of the proof, we obtain

$$(3.16) \qquad \prod_{j=2}^{k+1} (1 + \delta\delta(\eta)) \geq \frac{1}{2(1+\Theta)} \left\| \sum_{j=1}^{k+1} a_j y_j \right\| \geq$$

$$\geq \frac{1}{2(1+\Theta)} \frac{1}{3C_q} \left(\sum_{j=1}^{k+1} a_j^q \right)^{1/q} = \frac{1}{2(1+\Theta)} \frac{1}{3C_q} \frac{a_{k+2}}{\frac{1}{3C_q}\eta} =$$

$$= \frac{1}{2(1+\Theta)} \frac{1}{3C_q} \left[1 + \left(\frac{1}{3C_q} \eta \right)^q \right]^{k/q}.$$

By taking the k-th root from both sides of (3.16) we obtain

$$1+\delta\delta(\eta) \geq \left[\frac{1}{2(1+\Theta)} \frac{1}{3C_q} \right]^{1/k} \left[1 + \left(\frac{1}{3C_q} \eta \right)^q \right]^{1/q} \geq$$

$$\geq \left[\frac{1}{2(1+\Theta)} \frac{1}{3C_q} \right]^{1/k} \left[1 + \frac{1}{q} \left(\frac{1}{3C_q} \eta \right)^q \right].$$

Therefore

$$\underline{\lim}_{\eta\to 0} \frac{\delta\delta(\eta)}{\eta^q} \geq \frac{1}{q} \left(\frac{1}{3C_q} \right)^q.$$

This completes the proof of the Theorem.

For an infinite dimensional Banach space B, define

$$p_B = \sup \{p \mid B \text{ is of type } p\}$$

and

$$q_B = \inf \{q \mid B \text{ is of cotype } q\}.$$

It follows from (3.1) and (3.2) that

$$(3.17) \qquad \sup \{p \leq 2 \mid \overline{\lim}_{\eta\to 0} \beta\beta(\eta)/\eta^p = 0\} \geq p_B$$

and

$$(3.18) \qquad \inf \{q \geq 2 \mid \underline{\lim}_{\eta\to 0} \delta\delta(\eta)/\eta^q > 0\} \leq q_B.$$

By Maurey-Pisier Theorem [MaPi] (see [MSch], Chapter 13), formulas (3.17) and (3.18) imply Theorem 2.1. A well known fact has just to be added to the proof of part i) of Theorem 2.1 that, if ℓ_{p_0} is finitely representable in B for some $1 < p_0 < 2$, then, for any p, $p_0 < p \le 2$, ℓ_p is also finitely representable in B (see, e.g., [MSch], Chapter 8).

4. GEOMETRIC MODULI OF A BASIC SEQUENCE AND THEIR TYPE AND COTYPE.

In this section we assume that $X = \{x_j\}_1^\infty \subset B$ is a minimal system with the total dual $X^* = \{x_j^*\}_1^\infty$ and we consider the moduli $\beta\beta(\varepsilon) \equiv \beta\beta(\varepsilon, \mathcal{B}(X))$ and $\delta\delta(\varepsilon, \mathcal{B}(X)) \equiv \delta\delta(\varepsilon, \mathcal{B}(X))$. For $\Theta'>0$ we choose $E_{\Theta'} \in B(X)$ such that

$$\beta\beta(\varepsilon) \overset{\Theta'}{\simeq} \inf_{x \in S(E_{\Theta'})} \beta(\varepsilon; x; B(X))$$

and

$$\delta\delta(\varepsilon) \overset{\Theta'}{\simeq} \sup_{x \in S(E_{\Theta'})} \delta(\varepsilon; x; B(X)).$$

Denote by $Y = \{y_j\}_1^\infty$ a basic block sequence of X with $Y \subset E_{\Theta'}$, and such that Y satisfies the conclusions of Theorem 3.3. It is possible, as has been proved in [M], because $\mathcal{B}=B(X)$ has the so called filter property (i.e., for any E_1 and $E_2 \in \mathcal{B}$ there exists $E_3 \in \mathcal{B}$ such that $E_3 \subset (E_1 \cap E_2)$).

Let us begin with some definitions and notations. We say that a sequence $X \subset B$ is of block type p (block cotype q) if there exists a constant $C \ge 1$ ($D \ge 1$), such that for every positive integer n and for any n blocks $\left\{y_i = \sum_{j=n_i+1}^{n_{i+1}} a_j x_j\right\}_{i=1}^n$ of X, we have

$$\left[\int_0^1 \left\| \sum_{i=1}^n r_i(t)\, y_i \right\|^p dt\right]^{1/p} \le C\left[\sum_{i=1}^n \|y_i\|^p\right]^{1/p}$$

$$\left(D\left[\int_0^1 \left\| \sum_{i=1}^n r_i(t) y_i \right\|^q\right]^{1/q} \ge \left[\sum_{i=1}^n \|y_i\|^q\right]^{1/q}\right).$$

Here $r_i(t)$, $i = 1,2,\ldots$ is the sequence of Rademacher functions on $[0,1]$. The smallest C (resp. D) which satisfies the above inequality is called the block-type p constant and will be denoted by $T_p(X)$ (similarly, the block-cotype q constant and will be denoted by $C_q(X)$).

Remark. The constraints $1 \le p \le 2 \le q \le \infty$ which hold for type and cotype of a Banach space B do not hold in this case, and block type (block cotype) can be any value in $[1,\infty]$.

Denote

$$p_X = \sup\,\{p \,|\, p \text{ is a block type of } X\}$$

and

$$q_X = \inf\,\{q \,|\, q \text{ is a block type of } X\}.$$

Obviously $p_X \le q_X$. It is also obvious that for every block sequence X' of X, $p_X \le p_{X'} \le q_{X'} \le q_X$.

Proposition 4.1. *Let* Y *be the basic block sequence of* X *defined above*

(i) *If* $q_Y < \infty$ *then*

$$\varliminf_{\eta\to 0} \frac{\beta\beta(\eta; B(X))}{\eta^{q_Y}} \le C\cdot M^{q_Y},$$

where C *and* M *are absolute constants.*

(ii) *If* $q_Y = \infty$ *then there exists* $\varepsilon_0 > 0$ *such that* $\beta\beta(\varepsilon_0; B(X)) = 0$.

(iii) *If* $1 \le p_Y < \infty$ *then*

$$\overline{\lim_{\eta\to 0}} \frac{\delta\delta(\eta; B(X))}{\eta^{p_Y}} \ge (\ln 3) C_2^{p_Y},$$

where $C_2 > 0$ *is an absolute constant.*

Corollary 4.2. *Under the above notations*

(i) $$\sup\left\{q\ ;\ \underline{\lim_{\eta\to 0}} \frac{\beta\beta(\eta; B(X))}{\eta^q} = 0\right\} \ge q_Y$$

and

$$\inf\left\{p\ ;\ \overline{\lim_{\eta\to 0}} \frac{\delta\delta(\eta; B(X))}{\eta^q} > 0\right\} \le p_Y.$$

The proof of the corollary is obvious. The proof of the proposition relies on the following lemma.

Lemma A ([MSh1 and 2]). *There exists* $0 < \delta < 1$ *such that for every* $\varepsilon > 0$ *there exists a constant* $\gamma = \gamma(\varepsilon) > 0$ *with the following property: For every natural number* k *there exists a finite basic block sequence* $\{z_i\}_1^k$ *with respect to* Y *such that,*

$$1-\delta \le \|z_j\| \le 1, \quad j=1,\dots k;$$

the set $\{z_i\}_1^k$ *is an unconditional basic set with the unconditional constant* ≤ 3 *for every* k, *and* ε*-invariant to spreading (which means that for every* $s \le k$ *and* $1 \le i_1 < i_2 < \dots < i_s \le k$; $1 \le j_1 < j_2 < \dots < j_s \le k$, *and scalars* $\{\alpha_i\}_1^s$ *with* $|\alpha_i| \le 1$, $i=1,\dots,s$, *we have*

$$\left\|\sum_{m=1}^{s} \alpha_m z_{i_m}\right\| \le (1+\varepsilon)\left\|\sum_{m=1}^{s} \alpha_m z_{j_m}\right\|); \text{ also}$$

$$(4.1) \qquad \|\sum_{j=1}^{s} z_j\| \le \gamma s^{1/q_Y}, \quad s \le k.$$

Proof of Proposition 4.1. First assume that $q_Y < \infty$. Let k be a natural number. Assume that for $\{z_j\}_1^{k+1}$ the conclusion of Lemma A is satisfied. Denote $\hat{z}_j = z_j/\|z_j\|$, $j=1,\dots,k+1$. Using the left hand side inequality of (3.5) we obtain for $a_1=1$, and arbitrary $a_2,\dots,a_{k+1}$,

$$(1+\Theta)\prod_{j=2}^{k+1}\left[1+\beta\left(\frac{|a_j|}{\|\sum_{n=1}^{j-1} a_n\hat{z}_n\|}\,;\,\sum_{n=1}^{j-1} a_n\hat{z}_n\right)\right] \le \|\sum_{j=1}^{k+1} a_j\hat{z}_j\|.$$

Using Lemma A we infer that $\{z_j\}_1^{k+1}$ is the unconditional set with a constant ≤ 3 and that property (4.1) is fulfilled. Hence, for every $s \le$ $\le k+1$ we have

$$(4.2) \qquad \|\sum_{n=1}^{s} a_n\hat{z}_n\| \le 1+ \|\sum_{n=2}^{s} a_n\hat{z}_n\| \le 1+ \frac{3}{1-\delta} \max_{2\le n\le s}|a_n|\,\|\sum_{n=2}^{s} z_n\| \le$$

$$\le 1 + \frac{3\gamma}{1-\delta} \max_{2\le n\le s}|a_n|(s-1)^{1/q_Y}.$$

Let us define a finite sequence of scalars as follows: $a_1=1$, and $a_j=(1/k)^{1/q_Y}$, $j = 2,\dots,k+1$. Using (4.2) with these a_j's we obtain

$$\|\sum_{n=1}^{s} a_n\hat{z}_n\| \le 1 + \frac{3\gamma}{1-\delta}(1/k)^{1/q_Y}(s-1)^{1/q_Y} \le 1 + \frac{3\gamma}{1-\delta} \equiv M.$$

Thus we have

$$(4.3) \qquad \frac{a_j}{\|\sum_{n=1}^{j-1} a_n\hat{z}_n\|} \ge \frac{(1/k)^{1/q_Y}}{M}, \quad j\ge 2.$$

Using (4.3) and the monotonicity of $\beta(\varepsilon;x)$ as a function of ε we obtain as in the proof of Theorem 3.1 (taking Θ' small enough)

$$(4.4)\qquad \prod_{j=2}^{k+1}\left[1+\beta\beta\left(\frac{(1/k)^{1/q_Y}}{M}\right)\right] \le \frac{2}{1-\Theta}\left\|\sum_{n=1}^{k+1} a_n\hat{z}_n\right\| \le \frac{2M}{1-\Theta}\,.$$

Now, take the k-th root from both sides of (4.4); we have

$$(4.5)\qquad \beta\beta\left(\frac{(1/k)^{1/q_Y}}{M}\right) \le \left(\frac{2M}{1-\Theta}\right)^{1/k}-1 = \frac{1}{k}\ln\frac{2M}{1-\Theta} + o(1)\ (\text{for } k\to\infty).$$

Denote $\mu_k = \frac{(1/k)^{1/q_Y}}{M}$. Since (4.5) holds for every k and $\mu_k\to 0$ $(k\to\infty)$ we have, by (4.5),

$$(4.6)\qquad \varlimsup_{\mu\to 0}\frac{\beta\beta(\mu)}{\mu^{q_Y}} \le \ln\left(\frac{2M}{1-\Theta}\right) M^{q_Y}.$$

This proves the first part of Proposition 4.1.

Assume that $q_Y=\infty$. For a given natural k let $\{z_j\}_1^{k+1}$ be a finite sequence for which the conclusions of Lemma A hold. Using (4.1) we obtain

$$\left\|\sum_{j=1}^{s} z_j\right\| = \gamma,\quad s \le k+1.$$

For the sequence of scalars $\{a_j=1\}_{j=1}^{k+1}$ we obtain, as in the proof of the first part of Theorem 4.1,

$$(4.7)\qquad \beta\beta(1/\gamma) \le \left(\frac{2k}{1-\Theta}\right)^{1/k} - 1.$$

The last inequality holds for every natural k. Since the right hand side of (4.7) vanishes as k tends to infinity we obtain

$$\beta\beta(1/\gamma) = 0.$$

This proves the second part of Proposition 4.1. The proof of part (iii) of Proposition 4.1 also relies on Lemma A. However, the property (4.1) is replaced by the following property (see also [MSh1])

$$(4.8) \qquad \Big\| \sum_{j=1}^{s} z_j \Big\| \geq 1/\gamma \, s^{1/p_Y}, \quad s \leq k+1.$$

Remark. In [MSh1] it is shown that (4.8) holds with some $\gamma = \gamma(s)$ such that $\lim_{s\to\infty} \log \gamma(s)/\log s = 0$. We do not need this complication because, in our case, we start with the basic sequence $Y = \{y_j\}$ which automatically satisfies the condition dist $(y_n, \mathrm{span}\{y_j\}_1^{n-1}) \geq \delta > 0$ for $n\geq 2$. To organize this condition it was necessary in [MSh1] to reduce the low bound up to a factor $\gamma(s)^{-1}$.

Using the right hand side of inequality (3.5) we obtain

$$(1+\Theta) \prod_{j=2}^{k+1} \left[1+\delta\left(\frac{|a_j|}{\|\sum_{n=1}^{j-1} a_n \hat{z}_n\|}\, ; \sum_{n=1}^{j-1} a_n \hat{z}_n\right)\right] \leq \Big\| \sum_{n=1}^{k+1} a_n \hat{z}_n \Big\|.$$

Relying on the same arguments as in the proof of part (i) of this theorem we have

$$\Big\| \sum_{n=1}^{s} a_n \hat{z}_n \Big\| \geq \frac{1}{3} \min_{1\leq n\leq s} |a_n| \, \Big\| \sum_{n=1}^{s} \hat{z}_n \Big\| \geq \frac{1}{3} \min_{1\leq n\leq s} |a_n| \, \Big\| \sum_{n=1}^{s} z_n \Big\| \geq$$

$$\geq \frac{1}{3} \min_{1\leq n\leq s} |a_n| \, \frac{1}{\gamma} \, s^{1/p_Y}.$$

Let us define a finite set of scalars as follows: $a_1=1$, and for

$j = 2,\dots,k+1$, $a_j = 18(1+\Theta)\gamma(1/k)^{1/p_Y}$.

Since $\|\sum_{n=1}^{j} a_n\hat{z}_n\| \geq 1$, $j = 2,\dots,k+1$, we obtain

$$\frac{a_j}{\|\sum_{n=1}^{j-1} a_n\hat{z}_n\|} \leq a_j\ ,\quad j = 2,\dots,k+1.$$

Using the monotonicity of $\delta(\varepsilon;x)$ as a function of ε we obtain as in the previous part

$$\prod_{j=2}^{k+1}(1+\delta\delta(a_j)) \geq \frac{1}{2(1+\Theta)}\|\sum_{n=1}^{k+1} a_n\hat{z}_n\| \geq$$

$$\geq \frac{1}{2(1+\Theta)}\cdot 6(1+\Theta)\cdot\gamma(1/k)^{1/p_Y}\frac{1}{\gamma}(k+1)^{1/p_Y} \geq 3.$$

Denote $\mu_k = 18(1+\Theta)\cdot\gamma\cdot(1/k)^{1/p_Y}$. We have $(1+\delta\delta(\mu_k))^k \geq 3$ and therefore

$$\delta\delta(\mu_k) \geq 3^{1/k}-1. \tag{4.9}$$

As in the proof of part (i) of the theorem, (4.9) implies that

$$\overline{\lim_{\mu\to 0}}\ \frac{\delta\delta(\mu)}{\mu^{p_Y}} \geq \ln 3\left(\frac{1}{18\cdot(1+\Theta)\cdot\gamma}\right)^{p_y}. \ \square \tag{4.10}$$

The next results are a direct consequence of Proposition 4.1. We refer to [MSch], Chapter 11 and [MSh2] for the terminology used below.

Theorem 4.3. a) *Assume that ℓ_∞ is finitely represented by blocks in every basic block sequence* Y *with respect to* X. *Then there exists $\varepsilon_0 > 0$ such that $\beta\beta(\varepsilon_0;B(X)) = 0$.*

b) *Suppose that ℓ_1 is finitely represented by blocks in every basic block sequence* Y *with respect to* X. *Then there exist* $\varepsilon_1>0$ *and* $c>0$ *such that* $\delta\delta(\varepsilon;B(X)) \geq c\varepsilon$ *for every* $\varepsilon < \varepsilon_1$.

Proof: a) We infer by the assumption that $q_Y=\infty$. Thus using part (ii) of Proposition 4.1, there exists $\varepsilon_0 > 0$ with $\beta\beta(\varepsilon_0) = 0$.

b) As in a) we conclude that $p_Y=1$. So $\mu_k = c_1 \cdot \frac{1}{k}$ for some absolute constant c_1. Now using (4.8), we conclude that $\delta\delta(\mu_k;B(X)) \geq c_2\mu_k$. For $\mu_{k+1} < \varepsilon < \mu_k$ we obtain by the monotonicity of $\delta\delta(\varepsilon;B(X))$,

$$\frac{\delta\delta(\varepsilon;B(X))}{\varepsilon} \geq \frac{\delta\delta(\mu_{k+1};B(X))}{\mu_k} \geq c_3 \frac{k}{k+1} .$$

Thus $\delta\delta(\varepsilon;B(X)) \geq c\varepsilon$ for some absolute constant c and all $\varepsilon \leq \mu_1$. □

It is interesting to compare the last theorem with the known result from [M1] which we cited in section 2 (Theorem 2.A.b). Another consequence of Proposition 4.1 is the following result.

Theorem 4.4. a) *Assume that there exist* $q_0 \geq 1$, $c > 0$ *and* $\varepsilon_0 > 0$ *such that* $\beta\beta(\varepsilon,B(X)) \geq c\varepsilon^{q_0}$ *for every* $\varepsilon \leq \varepsilon_0$. *Then* ℓ_q *is finitely represented by blocks in* X *for some* $q \leq q_0$.

b) *If there exist* $p_0\geq1$ $c_1>0$ *and* $\varepsilon_1>0$ *such that* $\delta\delta(\varepsilon,B(X)) \leq c_1\varepsilon^{p_0}$ *for every* $\varepsilon \leq \varepsilon_1$, *then* ℓ_p *is finitely represented by blocks in* X *for some* $p \geq p_0$.

The proof is immediate from Proposition 4.1 and Maurey-Pisier's theorem [MaPi] (see [MSch] Chapter 13) as we have already used it in the end of section 3. Note only that we have to use the variant of Maurey-Pisier's theorem for blocks of a given sequence as it was done, for example, in [MSch2].

Finally, we are ready to prove Theorem 2.2, stated in section 2.

Proof of Theorem 2.2. First assume that $q_Y=\infty$, for a block sequence Y from Proposition 4.1. Then, using Proposition 4.1, there exists ε_0 such that $\beta\beta(\varepsilon_0,B(X)) = 0$ and by Maurey-Pisier's theorem ℓ_∞ is finitely represented by blocks of X.

Now assume that $1 \le p_Y \le q_Y < \infty$. Using (4.5), (4.8) and the monotonicity of $\beta\beta(\varepsilon)$ we obtain by the assumptions of the theorem

$$c_1 \frac{1}{k} \le \delta\delta(c_2(1/k)^{1/p_Y}) \le c\beta\beta(c_2(1/k)^{1/p_Y}) =$$
$$= c\beta\beta(c_2(1/k)^{1/p_Y - 1/q_Y}(1/k)^{1/q_Y}) \le c_3 \frac{1}{k}.$$

Denote $\mu_k = c_2(1/k)^{1/p_Y}$, $k=1,2,\dots$; then

$$A_1 \mu_k^{p_Y} \le \beta\beta(\mu_k) \le A_2 \mu_k^{p_Y}, \quad k=1,2,\dots .$$

For $\mu_{k+1} < \varepsilon < \mu_k$ we have

$$A_1 \left(\frac{\mu_{k+1}}{\mu_k}\right)^{p_Y} \le \frac{\beta\beta(\mu_{k+1})}{\mu_k^{p_Y}} \le \frac{\beta\beta(\varepsilon)}{\varepsilon^{p_y}} \le \frac{\beta\beta(\mu_k)}{\mu_{k+1}^{p_Y}} \le A_2 \left(\frac{\mu_k}{\mu_{k+1}}\right)^{p_Y}.$$

Because $\lim_{k\to\infty} \frac{\mu_{k+1}}{\mu_i} = 1$, we have, for $\varepsilon \to 0$

$$\beta\beta(\varepsilon, B(X)) = O(\varepsilon^{p_Y}) \quad \text{and} \quad \delta\delta(\varepsilon, B(X)) = O(\varepsilon^{p_Y}).$$

So, $p_Y = q_Y$ and, by Maurey-Pisier's theorem, we conclude that ℓ_{p_Y} is finitely represented by blocks of X. □

<u>Remark</u>. Clearly

$$\beta^0\beta^0(\varepsilon) \le \beta\beta(\varepsilon; B(X)) \le \delta\delta(\varepsilon; B(X)) \le \delta^0\delta^0(\varepsilon).$$

Therefore, if there exists $c \ge 1$ such that $\delta^0\delta^0(\varepsilon) \le c\beta^0\beta^0(\varepsilon)$ then the conclusion of Theorem 2.2 holds also for $\beta^0\beta^0(\varepsilon)$.

5. FINITE DIMENSIONAL MODULI $\beta\beta(\varepsilon,B_0)$ AND $\delta\delta(\varepsilon;B_0)$ AND THE LOCAL STRUCTURE OF A SPACE.

In this section, we consider the finite dimensional moduli $\beta\beta(\varepsilon;B_0) \equiv \beta_0\beta_0(\varepsilon;B_0)$ and $\delta\delta(\varepsilon;B_0)$ and compare their behavior with the type and cotype of the space in the spirit of section 3.

Theorem 5.1. *Let B be an infinite dimensional Banach space of type* $1\le p\le 2$ *and cotype* $2 \le q < \infty$. *Then*

$$\sup\left\{p;\ \overline{\lim_{\mu\ 0}}\ \frac{\delta_0\delta_0(\mu)}{\mu^p} = 0\right\} \ge P_B = \sup\{p \mid B \text{ is of type } p\},$$

and

$$\inf\left\{q;\ \underline{\lim_{\mu\ 0}}\ \frac{\beta_0\beta_0(\mu)}{\mu^q} > 0\right\} \le q_B = \inf\{q \mid B \text{ is of cotype } q\},$$

For the proof ot Theorem 5.1 we will need the following known fact (see [M1], Lemma, 5.1, or [M2], section 5.8).

Lemma A. *Let B be an infinite dimensional Banach space. There exists an integer function* $n(k;\Theta,C)$ *such that, for any n-dimensional subspace* $L_n \hookrightarrow B$ *and any* $x\in S(B)$, *there exists a k-dimensional subspace* $L_k \hookrightarrow L_n$ *and for any* ε, $0 \le \varepsilon \le C$,

$$\sup\left\{\|x+\varepsilon y\| \ \middle|\ y \in S(L_k)\right\} - \inf\left\{\|x+\varepsilon y\| \ \middle|\ y \in S(L_k)\right\} \le \Theta.$$

There exists, of course, a finite dimensional version of this lemma which obviously implies, by definition of moduli $\delta_k(\varepsilon;x)$ and $\beta_k(\varepsilon;x)$ (see section 1), the following proposition

Proposition 5.2. *Let B be an infinite dimensional Banach space, L an N-dimensional subspace of B and* L_n *an n-dimensional subspace of L. For* $C < \infty$, *a given* $\Theta > 0$ *and a positive integer m, there exists a positive integer* $N_0 = N(m,n,C,\Theta)$, *such that, if* $N > N_0$, *there exists a m-dimensional subspace* $L_m \subset L$ *such that for every* $x \in S(L_n)$, $y \in S(L_m)$ *and* $\varepsilon \in [0,C]$ *we have,*

$$(5.1) \qquad (1-\Theta)(1+\delta_m(\varepsilon;x)) \le \|x+\varepsilon y\| \le (1+\Theta)(1+\beta_m(\varepsilon;x)).$$

Remark. Use the proposition for $C = 1+2/\Theta$. Then, for $\varepsilon > C$, we have

$$(1+\Theta)\|x+\varepsilon y\| \ge (1+\Theta)(\varepsilon-1) \ge \left(1 + \frac{2}{\varepsilon-1}\right)(\varepsilon-1)=\varepsilon+1 \ge 1+\delta_k(\varepsilon;x).$$

Then

$$(1-\Theta)(1+\delta_k(\varepsilon;x)) \le (1-\Theta)(1+\Theta)\|x+\varepsilon y\| \le \|x+\varepsilon y\| \le$$
$$\le (1+\Theta)(1+\beta_k(\varepsilon;x)).$$

Therefore, (5.1) holds for every $\varepsilon > 0$.

We are now going to show a finite dimensional version of estimate (3.5).

Proposition 5.3. *Let* B *be an infinite dimensional Banach space. For all positive integers* m, k *and any* $\Theta > 0$, *there exists an integer* $N(m,k,\Theta)$ *such that in every* N*-dimensional subspace* $L \subset B$ *for* $N \ge N(m,k,\Theta)$ *we can select a set* $\{x_j\}_1^k \subset S(L)$ *such that for* $a_1=1$ *and* $a_j \in \mathbb{R}$, $2 \le j \le k$,

$$(5.2) \qquad (1-\Theta)\prod_{j=2}^{k}\left[1+\delta_m\left(\frac{|a_j|}{\|\sum_{n=1}^{j-1} a_n x_n\|};\sum_{n=1}^{j-1} a_n x_n\right)\right] \le \|\sum_{n=1}^{k} a_n x_n\| \le$$
$$\le (1+\Theta)\prod_{j=2}^{k}\left[1+\beta_m\left(\frac{|a_j|}{\|\sum_{n=1}^{j-1} a_n x_n\|};\sum_{n=1}^{j-1} a_n x_n\right)\right].$$

Moreover, we can replace $\{x_j\}_1^k$ *in (5.2) by every block set* $\{y_n\}_1^{\ell} \subset S(B)$ *of* $\{x_j\}_1^k$.

Proof. For a given $\Theta > 0$ let us select the numbers $\{\Theta_i > 0\}_{i=1}^{k-1}$, with $1+\Theta \ge \prod_{i=1}^{k-1}(1+\Theta_i)$ and $1-\Theta \le \prod_{i=1}^{k-1}(1-\Theta_i)$. Let $L_0 \subset B$ be an N_k-dimensional subspace with N_k large enough (we may compute this N_k as a k-iterated function of $N(m;N_{k-1};1+\frac{2}{\Theta_1};\Theta_1)$ where $N(m,n,C,\Theta)$ is defined in Proposition 5.2). Let $x_1 \in S(L_0)$. Using Proposition 5.2, we select a subspace $L_1 \subset L_0$, $\dim L_1 = N_{k-1}$, such that for every $y \in L_1$

$$(5.3) \qquad (1-\Theta_1)(1+\delta_m(\|y\|;x_1)) \le \|x_1+y\| \le (1+\Theta_1)(1+\beta_m(\|y\|;x_1)).$$

We continue by induction. Suppose that the elements $\{x_i\}_{i=1}^{n-1}$ were already constructed with the appropriate sequence of subspaces $L_{n-1} \subset L_{n-2}$. As in the first step, for $x_n \in S(L_{n-1})$ one can choose a subspace $L_n \subset L_{n-1}$, $\dim L_n = N_{k-n}$ and $N_{k-n+1} = N(m;N_{k-n};1+\frac{2}{\Theta_n};\Theta_n)$, such that for every $y \in L_n$ and $x \in S(\operatorname{span}\{x_j\}_1^n)$ we have

$$(5.4) \qquad (1-\Theta_n)(1+\delta_m(\|y\|;x)) \le \|x+y\| \le (1+\Theta_n)(1+\beta_m(\|y\|;x)).$$

Repeating the use of (5.4) we obtain, for $a_1=1$,

$$\Big\|\sum_{j=1}^{k} a_jx_j\Big\| = \Big\|\sum_{j=1}^{k-1} a_jx_j\Big\| \left\| \frac{\sum_{j=1}^{k-1} a_jx_j}{\|\sum_{j=1}^{k-1} a_jx_j\|} + \frac{a_kx_k}{\|\sum_{j=1}^{k-1} a_jx_j\|} \right\| \le$$

$$\le (1+\Theta_{k-1})\Big\|\sum_{j=1}^{k-1} a_jx_j\Big\| \left[1+\beta_m\left(\frac{|a_k|}{\|\sum_{j=1}^{k-1} a_jx_j\|};\sum_{j=1}^{k-1} a_jx_j\right)\right] \le \dots \le$$

$$\le \prod_{j=1}^{k-1}(1+\Theta_j)\prod_{j=2}^{k}\left[1+\beta_m\left(\frac{|a_j|}{\|\sum_{n=1}^{j-1} a_nx_n\|};\sum_{n=1}^{j-1} a_nx_n\right)\right] \le$$

$$\le (1+\Theta)\prod_{j=2}^{k}\left[1+\beta_m\left(\frac{|a_j|}{\|\sum_{n=1}^{j-1} a_nx_n\|};\sum_{n=1}^{j-1} a_nx_n\right)\right].$$

This proves the right hand side of (5.2). In a similar way we prove the left hand side of (5.2).

Note that by construction we may replace $\{x_j\}_1^k$ in (5.2) by every block sequence $\left\{y_\ell = \sum_{j=n_\ell+1}^{n_{\ell+1}} a_j x_j\right\}_{\ell=1}^{p}$, of $\{x_j\}_1^k$ (since the sequence of subspaces $\{L_j\}_0^k$ is a decreasing sequence and span $\{y_\ell\}_{\ell=1}^{s} \subset$ span $\{x_j\}_1^{n_{s+1}}$).

We return to the proof of Theorem 5.1. Let $\eta, \kappa > 0$ be given. There exists a positive integer m such that

$$(5.5)\qquad |\delta_m \delta_m(\eta) - \delta_0 \delta_0(\eta)| < \kappa$$

and

$$|\beta_m \beta_m(\eta) - \beta_0 \beta_0(\eta)| < \kappa.$$

The functions $\beta_m(\lambda;x)$ and $\delta_m(\lambda;x)$ are Lipshitz on $[0,1] \times S(B)$. Then, by [M1] (Theorem 5.1), for any N, there exists an N-dimensional subspace $L_0 \subset B$ such that for every $\lambda \in [0,1]$ the oscillation of the two fuctions $\beta_m(\lambda;x)$ and $\delta_m(\lambda;x)$ on $S(L_0)$ is small (say, less than a given $\varepsilon > 0$). Use Proposition 5.3 with this subspace L_0, then

$$(5.6)\qquad 1+\delta_m\left[\eta;\ \sum_{n=1}^{j-1} a_n x_n\right] \ge (1-\varepsilon)(1+\sup_{x\in S(L_0)} \delta_m(\eta;x)) \ge (1-\varepsilon)(1+\delta_m\delta_m(\eta)).$$

Suppose that for a given k we select ε in such a way that

$$(1-\varepsilon)^k \ge 1/2.$$

Using similar arguments to those in section 3 together with the left hand side of (5.2) and (5.6) we obtain

$$(5.7)\qquad \prod_{j=2}^{t_k+1} (1+\delta_m\delta_m(\eta)) \le \left[\frac{2}{1-\Theta}\, 3T_p\right](1+(3T_p\eta)^p)^{t_k/p}.$$

for some integer function t_k such that $t_k \longrightarrow \infty$ (if $k \longrightarrow \infty$).

Assume that κ has been chosen in such a way that

$$(1-\kappa)^k \geq 1/2.$$

Using (5.5) and (5.7) we have

$$(1+\delta_0\delta_0(\eta))^{t_k} \leq \left[\frac{4}{1-\Theta}\, 3T_p\right] (1+(3T_p\eta)^p)^{t_k/p}.$$

Taking the t_k-th root from both sides of the above inequality we obtain (for $t=t_k$)

$$1+\delta_0\delta_0(\eta) \leq \left[\frac{4}{1-\Theta}\, 3T_p\right]^{1/t} (1+(3T_p\eta)^p)^{1/p} \leq$$

$$\leq \left[\frac{4}{1-\Theta}\, 3T_p\right]^{1/t} (1+(3T_p\eta)^p).$$

Let $t\longrightarrow\infty$, then

$$(5.8) \qquad \delta_0\delta_0(\eta) \leq (3T_p\eta)^p$$

and

$$\overline{\lim_{\eta\to 0}} \frac{\delta_0\delta_0(\eta)}{\eta^p} \leq (3T_p)^p.$$

The proof of the formula for $\beta_0\beta_0(\varepsilon)$ is similar. We obtain

$$(5.9) \qquad \frac{\beta_0\beta_0(\mu)}{\mu^q} \geq 1/q \left[\frac{1}{3C_q}\right]^q$$

and then take the lim inf. □

As a corollary of Theorem 5.1 (or, to be more exact, from (5.8) and (5.9)), we obtain Theorem 2.3 stated in section 2 in the same way as we proved Theorem 2.1 at the end of section 3.

REFERENCES.

[Cl] Clarkson, J.A. (1962). Orthogonality in normed linear spaces. Archiv Math. 4:4, 297-318.

[D] Day, M.M. (1941). Reflexive Banach spaces not isomorphic to uniformly convex spaces. Bull. Amer. Math. Soc. 47:4, 313-317.

[K] Kadeč, M.I. (1956). Unconditionally convergent series in a uniformly convex space. Uspekhi Mat. Nauk 11:5, 185-190 (Russian).

[Li] Lindenstrauss, J. (1963). On the modulus of smoothness and divergent series in Banach spaces. Michigan Math. J. 10:3, 241-252.

[M1] Milman, V. (1971). Geometric theory of Banach spaces II. Geometry of the unit sphere. Uspechi Mat. Nauk, 26 no. 6, 73-149; Russian Math. Surveys 26, 6, 80-159.

[M2] Milman, V.D. (To appear),The heritage of P. Levy in geometric functional analysis. Asterisque.

[M3] Milman, V.D. (1967). Infinite-dimensional geometry of the unit sphere of a Banach space. Soviet Math. Dokl. 8, 1440-1444 (Translated from Russian).

[MaPi] Maurey, B. & Pisier, G. (1976). Series des variables aleatoires vectorielles independantes et priorietes geometriques des espaces de Banach. Studia Math. 58, 45-90.

[Mi1] Milman, D.P. (1938). Certain tests for the regularity of spaces of type B. Dokl. Akad. Nauk SSSR 20, 243-246 (Russian).

[MSch] Milman, V.D. & Schechtman, G. (1986). Asymptotic theory of finite dimensional normed spaces. Springer-Verlag. Lecture Notes in Mathematics 1200, 156 pp.

[MSh1] Milman, V.D. & Sharir, M. (1979). A new proof of the Maurey-Pisier theorem. Israel J. of Math. 33, 1, 73-87.

[MSh2] Milman, V.D. & Sharir, M. (1979). Shrinking minimal systems and complementation of n_p-spaces in reflexive Banach spaces. Proceeding of London Math. Soc., (3) 39, 1-29.

[Sm] Shmul'yan, V.L. (1940). On the differentiability of the norm of a Banach space. Dokl. Acad. Nauk SSSR 27, 643-648.

HILBERT SPACE REVISITED.

J. R. Retherford
Louisiana State University
Baton Rouge, LA 70803

Dedicated to Professor Antonio Plans

1. INTRODUCTION.

Since some of the work of A. Plans concerns operators on Hilbert spaces (e.g. [Pl 1-5], [PB]) it seems appropriate, in this "Festschrift" on the occasion of the retirement of Professor Plans, to look back at a few (of many) operator characterizations of Hilbert space. We will show that <u>all</u> of these characterizations follow easily from the summability property of the eigenvalues of nuclear operators on isomorphs of Hilbert space. This property is discussed below.

2. DEFINITIONS.

Recall that an operator T from a Banach space X to a Banach space Y, here after written $T : X \longrightarrow Y$, is <u>nuclear</u> provided there are (f_n) in X^*, (y_n) in Y such that

$$Tx = \sum_{n=1}^{\infty} f_n(x)\, y_n$$

for each $x \in X$, and

$$\sum_{n=1}^{\infty} \|f_n\|\, \|y_n\| < +\infty.$$

This concept is due to Grothendieck [G3] and Ruston [Ru].

On Hilbert spaces the nuclear operators coincide with the "trace-class" operators $\mathscr{S}_1$, where $T \in \mathscr{S}_1$ means

$$\sum_{n=1}^{\infty} |\lambda_n(\sqrt{TT^*})| < +\infty.$$

Here $(\lambda_n\sqrt{TT^*})$ denotes the eigenvalues of the unique positive square root of TT^*.

Moreover, it is readily seen that

$$\mathcal{S}_1 = \mathcal{S}_2 \cdot \mathcal{S}_2$$

where $\mathcal{S}_2$ denotes the Hilbert-Schmidt class. That is, each nuclear operator on a Hilbert space is the composition of two Hilbert-Schmidt maps and, conversely, the composition of two such maps is nuclear.

The generalization of this fact to Banach spaces (with applications) is the subject of this paper.

The Hilbert-Schmidt operators have been generalized to Banach spaces in the following way [G1], [Pi3], [S]: An operator $T: X \longrightarrow Y$ is absolutely two summing, written $T \in \Pi_2(X,Y)$, provided there is a constant K such that

$$\left(\sum_{i=1}^{n} \|Tx_i\|^2\right)^{1/2} \leq K \left(\sup_{\|f\|=1} \sum_{i=1}^{n} |f(x_i)|^2\right)^{1/2}$$

for all finite sets $x_1, \ldots, x_n$ in X.

Pietsch [Pi3] has shown that for Hilbert spaces H_1, H_2

$$\Pi_2(H_1, H_2) = \mathcal{S}_2(H_1, H_2)$$

so Π_2 is a natural generalization of the Hilbert-Schdmidt class.

The class Π_p, defined by replacing 2 by p, $1 \leq p < +\infty$, above, has seen numerous applications in Banach space theory.

The famous Grothendieck-Pietsch factorization theorem [G1] [Pi2,5] asserts that $T \in \Pi_2(X,Y)$ if and only if T admits the factorization

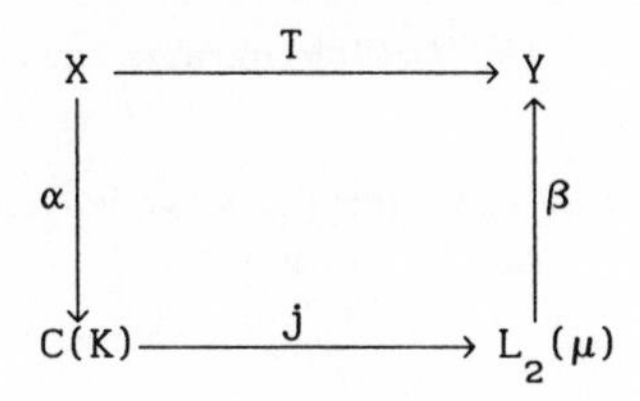

where K is a compact Hausdorff space, μ a normalized Borel measure on K, and j the inclusion mapping. It is easily seen that $j \in \Pi_2(C(K), L_2(\mu))$.

We will also have occasion to use the p-nuclear operators, in what follows: T is p-nuclear, $1 \le p < +\infty$, written $T \in N_p(X,Y)$ if

$$Tx = \sum_{n=1}^{\infty} f_n(x) y_n$$

(f_n) in X^*, (y_n) in Y, x in X where

$$\sum_{n=1}^{\infty} \|f_n\|^p < +\infty \quad \text{and} \quad \sup_{\|f\|=1} \sum_{n=1}^{\infty} |f(y_n)|^q < +\infty , \quad \frac{1}{p} + \frac{1}{q} = 1.$$

In particular 1-nuclear operators are just the nuclear operators.

3. EIGENVALUES OF NUCLEAR OPERATORS ON HILBERT SPACES.

Since we will be concerned with eigenvalues, we assume all our spaces to be complex Banach spaces. Let H be a Hilbert space and let $T \in N(H) = N(H,H)$ have eigenvalues $(\lambda_n(T))$ (ordered by decreasing modulus and counting multiplicities). The classical Weyl inequality [W] asserts that

$$\sum_{n=1}^{\infty} |\lambda_n(T)| \le \sum_{n=1}^{\infty} |\lambda_n \sqrt{TT^*}|$$

so the eigenvalues of an arbitrary nuclear operator are absolutely convergent. Surprisingly, Hilbert space is the <u>only</u> Banach space enjoying this property:

Theorem [JKMR]. *Let X be a Banach space. The following are equivalent:*

(a) X *is isomorphic to a Hilbert space;*

(b) *each nuclear operator on* X *has absolutely summing eigenvalues; and*

(c) $N(X) = \Pi_2 \cdot \Pi_2(X)$.

Since (c) is critical to our work we discuss it in some detail. Firstly, the notation $T \in \Pi_2 \cdot \Pi_2(X)$ means there is a Banach space Y and operators $U \in \Pi_2(X,Y)$, $V \in \Pi_2(Y,X)$ with $T = VU$.

Secondly, if $T \in \Pi_2 \cdot \Pi_2(X)$ then T has absolutely summing eigenvalues. To see this, following Pietsch [Pi5] we say that A and B are related operators if

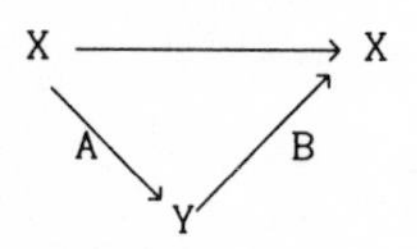

for Banach spaces X and Y.

The importance of this notion is that AB and BA have, in this case, the same eigenvalues. Thus if $T \in \Pi_2 \cdot \Pi_2(X)$ we have for some Banach space Y, $T = BA$ and $A \in \Pi_2(X,Y)$, $B \in \Pi_2(Y,X)$. By Grothendieck-Pietsch factorization we thus have the following diagram:

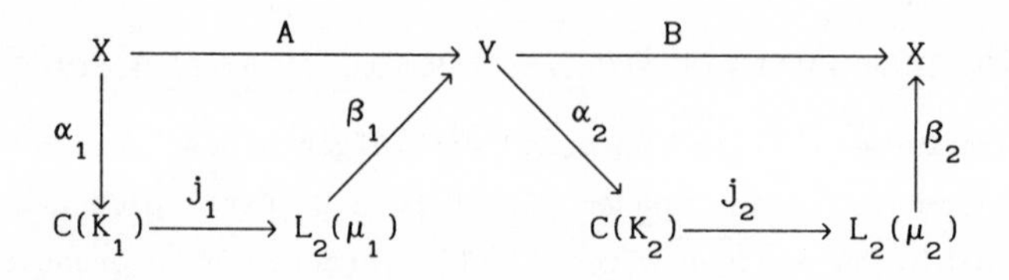

Now $(j_1\alpha_1\beta_2)(j_2\alpha_2\beta_1) \in \mathscr{S}_2 \cdot \mathscr{S}_2(L_2(\mu_1)) = \mathscr{S}_1(L_2(\mu_1))$ and so this operator has, by the Weyl inequality, absolutely summing eigenvalues. Clearly this operator is a relative of BA, i.e. T has absolutely summing eigenvalus. Our strategy is thus the following:

> *to show that a Banach space* X *is isomorphic to a Hilbert space we need only show that for* $T \in N(X)$, *the eigenvalues of* T, $(\lambda_n(T))$, *satisfy*

$$\sum_{n=1}^{\infty} |\lambda_n(T)| < +\infty.$$

To accomplish this we show, under various hypotheses, that T (or T^*) is in $\Pi_2 \cdot \Pi_2$.

To do this we need the "big picture" provided by à famous result of Grothendieck.

4. GROTHENDIECK FACTORIZATION.

As usual we denote by ℓ_p, $1 \le p < +\infty$ the Banach space of sequences $\lambda=(\lambda_n)$ satisfying $\|\lambda\|_p = \left(\sum_{n=1}^{\infty} |\lambda_n|^p\right)^{1/p} < +\infty$, and by ℓ_∞ the space of sequences $\lambda = (\lambda_n)$ with $\|\lambda\|_\infty = \sup_n |\lambda_n| < +\infty$.

Finally c_0 denotes the closed subspace of ℓ_∞ consisting of those sequences $\lambda = (\lambda_n)$ with $\lim_{n\to\infty} \lambda_n = 0$.

Let $f \otimes y$, $f \in X^*$, $y \in Y$ denote the rank one operator $x \quad f(x)y$. Thus any non-zero $T \in N(X)$ can be written

$$T = \sum_{n=1}^{\infty} f_n \otimes y_n \quad \text{with} \quad \sum_{n=1}^{\infty} \|f_n\| \, \|y_n\| < +\infty$$

and $f_n \ne 0$, $y_n \ne 0$ for each n. Let $\lambda_n = \|f_n\| \, \|y_n\|$ and write $\lambda_n = \alpha_n \beta_n$ where, $\alpha_n > 0$, $(\alpha_n) \in c_0$ and $(\beta_n) \in \ell_1$. Then

$$f_n \otimes y_n = \beta_n \alpha_n^{1/2} \frac{f_n}{\|f_n\|} \otimes \alpha_n^{1/2} \frac{y_n}{\|y_n\|} = \beta_n g_n \otimes z_n$$

clearly $(\|g_n\|)$, $(\|z_n\|) \in c_0$ and

$$T = \sum_{n=1}^{\infty} \beta_n g_n \otimes z_n.$$

This yields the Grothendieck factorization theorem for nuclear operators

[G1]: if $T \in N(X,Y)$, then T admits the following factorization:

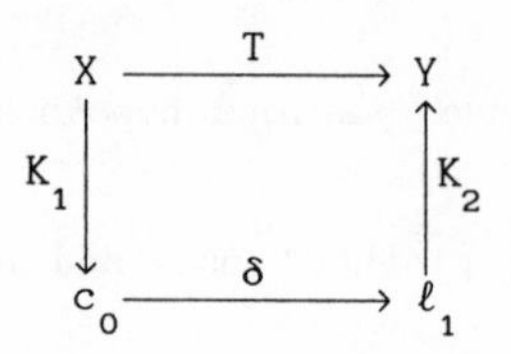

where K_1 and K_2 are compact and δ is a diagonal nuclear operator. Indeed let (β_n), (g_n), (z_n) be as above and let $K_1(x) = (g_n(x)$, $K_2((\xi_n)) = \sum_{n=1}^{\infty} \xi_n z_n$ and $\delta((\xi_n)) = (\beta_n \xi_n)$.

Observe that δ admits a further factorization

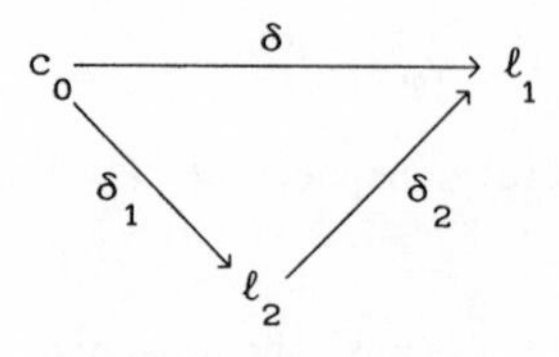

where δ_1 and δ_2 are the diagonal mappings defined by $(\beta_n^{1/2})$. It is easy to show from the definition that $\delta_1 \in \Pi_2(c_0, \ell_2)$ (even $N_2(c_0, \ell_2)$) and $\delta_2^* \in \pi_2(\ell_\infty, \ell_2)$. Or, if one desires, a deep result of Grothendieck [G1] also covers the above situation.

This diagram is now produced in it's entirety. Our strategy outlined above reduces to the intellectual (?) game -chasing the diagram!

Grothendieck Factorization for $T \in N(X)$

$K_1, K_2, \delta, \delta_1, \delta_2$ have the meanings as above.

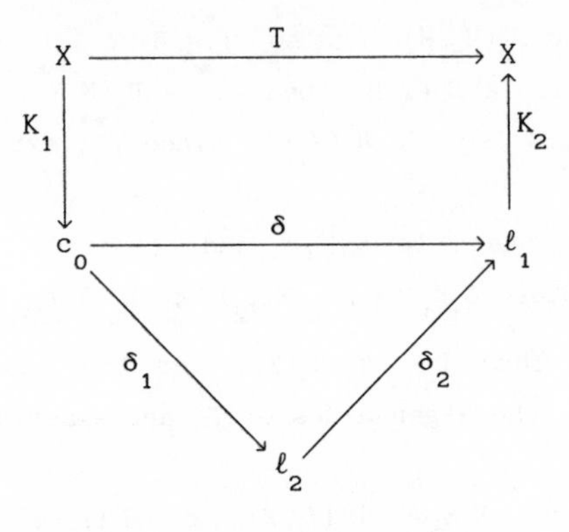

Main observation (Related operators): $(\delta_1K_1)(K_2\delta_2)$ has the same eigenvalues as T.

Actually, starting anywhere in the diagram and going all the way around yields an operator with eigenvalues the same as T. For example, $K_1K_2\delta$ is an operator on c_0 with eigenvalues the same as T.

5. APPLICATIONS.

In what follows we will introduce classes of spaces and operators as needed. We will write $X \in H$ to mean X is isomorphic to a Hilbert space.

I. Theorem of Cohen-Kwapien [C], [K]:

An operator $T \in D_2(X,Y)$ provided $T^* \in \Pi_2(Y^*,X^*)$.

Theorem: *Let* X *be a Banach space. The following are equivalent:*

(i) $X \in H$;

(ii) $\Pi_2(X,Y) \subset D_2(X,Y)$ *for every Banach space* Y;

(iii) $D_2(Y,X) \subset \Pi_2(Y,X)$ *for every Banach space* Y; *and*

(iv) *Same as* (i) *or* (iii) *with* $Y = \ell_2$.

Proof: (i)⇒(ii). If X is isomorphic to a Hilbert space H and $T \in \Pi_2(H,Y)$, by Grothendieck-Pietsch factorization we have

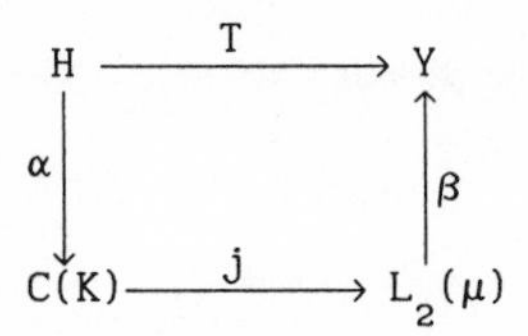

and $j\alpha \in \mathcal{S}_2(H,H_0)$ so $\alpha^* j^* \in \mathcal{S}_2(H_0,H)$ and so $T^* \in \Pi_2(Y^*,H)$.
(i)⇒(iii) is very similar: If $T \in D_2(Y,H)$ then $T^* \in \Pi_2(H,Y^*)$ so by what was just shown $T^{**} \in \Pi_2(Y^{**},H)$ so $T \in \Pi_2(Y,H)$ since T^{**} extends T with Y canonically imbedded in Y^{**}.
(ii)⇒(i) and (iii)⇒(i): Let $T \in N(X)$ and consider Grothendieck factorization. If (ii) holds $\delta_1 K_1 \in \Pi_2(X,\ell_2) \subset D_2(X,\ell_2)$. Also $\delta_2 \in D_2(\ell_2,\ell_1)$ so $T \in D_2 \cdot D_2(X)$. Thus $T^* \in \Pi_2 \cdot \Pi_2(X^*)$ and thus has absolutely converging eigenvalues. But, the eigenvalues of T^* are exactly those of T, so $X \in H$.

If (iii) holds $K_2\delta_2 \in D_2(\ell_2,X) \subset \Pi_2(\ell_2,X)$ and so $(K_2\delta_2)(\delta_1 K_1) \in \Pi_2 \cdot \Pi_2(X)$ and T has absolutely converging eigenvalues.
(iv)⇒(i) is clear since all we used above was the space ℓ_2.

An isometric version of I is valid. Indeed isometric results are valid for most of the theorems we give. To state such results requires conditions on the various norms of the ideals involved. Since we have not explicitly given these norms we will not give the isometric results.

II. Theorem of Lindenstrauss-Pelczynski [LP]:

One of the remarkable results of Grothendieck is the fact that $\mathcal{L}(L_1,H)$, the space of all bounded linear operators from an $L_1(\mu)$-space to a Hilbert space, coincides with $\Pi_1(L_1,H)$. This result is proved using what is now called the Grothendieck inequality [G2], [LP].

Theorem. *If X has an unconditional basis and $\mathcal{L}(X,Y) = \Pi_1(X,Y)$ then X is isomorphic to ℓ_1 and Y is isomorphic to a Hilbert space.*

Proof: Let (x_n,f_n) be an unconditional basis for X and assume $\|x\|=1$, $\|f_n\| \le M$. Let $\sup_{\|x\|=\|f\|=1} \Sigma|f_n(x)f(x_n)| = K < +\infty$. Let $(\alpha_n) \in c_0$, $0 < \alpha_n < 1$, $\alpha_{n+1} < \alpha_n$ and choose (y_n) in Y, $\|y_n\| = \alpha_n$ with $\sup_{\|g\|=1} \Sigma|g(y_n)|^2 \le 2$. (see e.g. [MR] for such a construction).

First, let $(\beta_n) \in \ell_2$, $\|(\beta_n)\|_2 = 1$. For each $N \ge 1$, let

$T_N x = \sum_{n=1}^{n} \beta_i f_i(x) y_i$. Then

$$\|T_N x\| \leq \sup_{\|g\|=1} \sum_{i=1}^{n} |f_i(x)\|\beta_i\|g(y_i)|$$

$$\leq M\|x\|(\Sigma|\beta_i|^2)^{1/2}(\sup_{\|g\|=1} \Sigma|g(y_i)|^2)^{1/2} \leq 2M\|x\|$$

so each T_N is continuous. By hypothesis there is a constant c, such that

$$\sum_{n=1}^{N} \|T_N(f_i(x)x_i)\| = \sum_{i=1}^{N} \|\beta_i f_i(x)y_i\| = \sum_{i=1}^{N} |f_i(x)||\beta_i||\alpha_i|$$

$$\leq c \sup_{\|f\|=1} \sum_{n=1}^{N} |f_i(x)f(x_i)| \leq c\, K\|x\|.$$

Since (β_n) was arbitrarily chosen from the unit ball of ℓ_2, $(f_i(x))\in\ell_2$. Knowing $(f_i(x))\in \ell_2$ for $x \in X$, allows us to repeat the construction above <u>without</u> the sequence (β_n). Since $(\alpha_n)\in c_0$ (being positive and monotone doesn't matter) and $\|y_n\| = \alpha_n$ we now obtain $(f_i(x))\in \ell_1$. Since $\|x_i\| = 1$ it follows that X is isomorphic to ℓ_1.

Now let $T \in N(Y)$ and go to the diagrams. By hypothesis (since X is isomorphic to ℓ_1)

$$K_2 \in \Pi_1(\ell_1, Y) \subset \Pi_2(\ell_1, Y).$$

Thus

$$(K_2\delta_2)(\delta_1 K_1) \in \Pi_2 \cdot \Pi_2(Y)$$

so T has absolutely summing eigenvalues.

A word of explanation is in order. While our proof avoids the Grothendieck inequality, the construction used requires the Dvoretzky theorem [D1]. Moreover the proof of the eigenvalue result we are using requires the deep result of Lindenstrauss-Tzafriri [LT] that a space is isomorphic to a Hilbert space if and only if each of its closed subspaces is complemented. This result in turn uses [D2].

So we've replaced <u>one</u> deep result with, essentially, <u>two</u> equally deep results! Still, diagram chasing's fun. Let's continue.

III. Theorem of Morrell-Retherford [MR]:

We write $T \in \Gamma_2(X,Y)$ if there is a Hilbert space H and operators $A \in \mathcal{L}(X,H)$, $B \in \mathcal{L}(H,Y)$ such that $T = BA$, i.e., T factors through a Hilbert space.

A Banach space X is in the class $\mathcal{D}_\infty$ if there is a constant such that for each n there is a subspace X_n of X and an isomorphism $\phi_n : X_n \longrightarrow \ell_\infty^n$ (complex n-space with the sup-norm) such that

$$\|\phi_n\|\|\phi_n^{-1}\| \le C.$$

Theorem: *The following are equivalent for a Banach space* X :

(i) $X \in H$;

(ii) $\mathcal{L}(X,Y) \subset \Gamma_2(X,Y)$ *for any* $Y \in \mathcal{D}_\infty$;

(iii) $\mathcal{L}(X,c_0) \subset \Gamma_2(X,c_0)$.

Proof: (i)⇒(ii)⇒(iii) is trivial (as is it all).

(iii)⇒(i). Diagraming again, we have by hypothesis that $K_1 \in \Gamma_2(X,c_0)$ so there is a Hilbert space H with

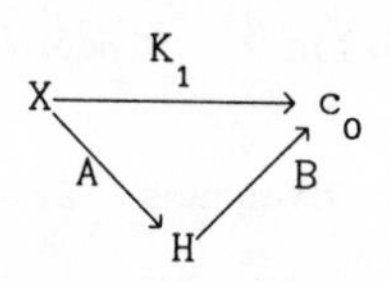

Then $AK_2\delta B \in N(H)$ and as before $X \in H$.

This shows we may remove the "p" in the title of [MR].

IV. Theorem of Gordon-Lewis-Retherford [GLR]:

Here C(X,Y) denotes the closure of the finite rank operators from X to Y.

Theorem: *The following are equivalent*:

(i) $X \in H$;

(ii) $C(X,Y) \subset \Gamma_2(X,Y)$ *for all* Y;

(iii) $C(X,c_0) \subset \Gamma_2(X,c_0)$;

(iv) $C(Y,X) \subset \Gamma_2(Y,X)$ *for all* Y; *and*

(v) $C(\ell_1,X) \subset \Gamma_2(\ell_1,X)$.

Proof. That (i)⇒(ii)⇒(iii) and (i)⇒(iv)⇒(v) is trivial.

(iii)⇒(i). Let $T \in N(X)$ and proceed to the diagram. By hypothesis $K_1 \in \Gamma_2(X, c_0)$ so there are operators A, B to and from a Hilbert space H, with $K_1 = BA$.

Now $\delta_2 B \in \mathcal{S}_2(H, \ell_2)$ and so also to $D_2(H, \ell_2)$, i.e. $T \in D_2 \cdot D_2(X)$.

(iv)⇒(i). Again let $T \in N(X)$. Now $K_2 \in \Gamma_2(\ell_1, X)$ and thus A, B, H exist with $K_2 = BA$ factoring through H. Hence $A\delta_2 \in D_2(\ell_2, H)$ and so to $\Pi_2(\ell_2, H)$. Thus $T \in \Pi_2 \cdot \Pi_2(X)$.

In [GLR] we also have characterizations of Hilbert space in terms of operators on X^*. These results can be obtained as above by dualizing the diagram.

V. Theorem of Holub [H]:

An operator T is quasi-p-nuclear, $1 \le p < +\infty$, $T \in QN_p(X, Y)$, if there is a sequence (f_n) in X^*, $\sum_{n=1}^{\infty} \|f_n\|^p < +\infty$ such that

$$\|Tx\| \le \left(\sum_{n=1}^{\infty} |f_i(x)|^p \right)^{1/p}$$

for each $x \in X$. This is equivalent to saying that if $i: Y \longrightarrow \ell_\infty(\Gamma)$ is an isometric imbedding of Y into a suitable ℓ_∞-space (the index set Γ may be uncountable) then $iT \in N_p(X, \ell_\infty(\Gamma))$.

Theorem: *The following are equivalent:*

(i) $X \in H$;

(ii) $T \in N_2(X, \ell_2) \Rightarrow T^* \in N_2(\ell_2, X^*)$;

(iii) $T \in N_2(X, \ell_2) \Rightarrow T^* \in QN_2(\ell_2, X^*)$;

(iv) $T \in QN_2(X, \ell_2) \Rightarrow T^* \in N_2(\ell_2, X^*)$;

(v) $T \in QN_2(X, \ell_2) \Rightarrow T^* \in QN_2(\ell_2, X^*)$.

Proof: Clearly $N_2 \subset QN_2 \subset \Pi_2$ regardless of the spaces involved and since for Hilbert spaces H_1, H_2, $\Pi_2(H_1, H_2) = \mathcal{S}_2(H_1, H_2)$, a glance at the Schmidt representation shows that

$$\Pi_2(H_1, H_2) = N_2(H_1, H_2).$$

Thus,

$$\mathscr{S}_2(H_1,H_2) = N_2(H_1,H_2) = QN_2(H_1,H_2) = \Pi_2(H_1,H_2)$$

for Hilbert spaces. So we certainly have (i)⇒(ii)⇒(iii) and (i)⇒(iv)⇒(v)⇒(iii). Thus we need only show (iii) ⇒ (i). Let $T \in N(X)$. Rushing once again to the diagram we see that by construction, $\delta_1 \in N_2(c_0,\ell_2)$ so by (iii) $K_1^*\delta_1^* \in QN_2(\ell_2,X^*) \subset \Pi_2(\ell_2,X^*)$. Also $\delta_2^*K_2^* \in \Pi_2(X^*,\ell_2)$, i.e. $T^* \in \Pi_2\cdot\Pi_2(X^*)$ so T has absolutely converging eigenvalues.

VI. Theorem of Rosenberger [R]:

We end these Hilbert space characterizations with a result which requires a generalization of the Hilbert-Schmidt operators to a class of operators <u>different</u> from the absolutely two summing operators. For $T \in \mathscr{L}(X,Y)$ let

$$\alpha_n(T) = \inf\{\|T-A\|: \text{rank } A \le n-1\}.$$

On Hilbert space $\alpha_n(T) = \lambda_n(\sqrt{TT^*})$ [Pi4].

Let, for $1 \le p < +\infty$,

$$S_p(X,Y) = \left\{T \in \mathscr{L}(X,Y) \mid \sum_{n=1}^{\infty} \alpha_n^p(T) < +\infty\right\}$$

Then clearly for Hilbert spaces H_1, H_2

$$S_2(H_1,H_2) = \mathscr{S}_2(H_1,H_2).$$

However, for arbitrary Banach spaces X,Y it is not true that $\Pi_2(X,Y) = S_2(X,Y)$. These ideas are due to Pietsch [Pi4]. Using multiplicative properties of $\alpha_n(T)$ it is easy to show that

$$S_2 \cdot S_2 \subset S_1.$$

Also [JKMR] if $T \in S_1(X)$ then the eigenvalues of T are absolutely summable. Thus (same idea as the strategy) to show that $X \in H$ it is enough to show that $T \in N(X)$ implies (under various hypotheses) that $T \in S_2 \cdot S_2(X)$.

Theorem: *The following are equivalent:*

(i) $X \in H$;

(ii) *for all* p, $1 \le p < \infty$, $N_p(X,\ell_2) \subset S_2(X,\ell_2)$ *and* $N_p(X^*,\ell_2) \subset S_2(X^*,\ell_2)$;

(iii) *there is a* p, $2 \le p < +\infty$ *such that* $N_p(X,\ell_2) \subset S_2(X,\ell_2)$ *and* $N_p(X^*,\ell_2) \subset S_2(X^*,\ell_2)$.

Proof: i)⇒(ii). It is easily seen that $N_p(H_1,H_2) \subset \Pi_p(H_1,H_2)$ for Hilbert spaces H_1, H_2. A. Pelczynski [P] has shown that in this case

$$\Pi_p(H_1,H_2) = \Pi_2(H_1,H_2)$$

for all p, $1 \le p < +\infty$. Thus if $p \ge 2$

$$\mathcal{S}_2(H_1,H_2) = \Pi_2(H_1,H_2) \subset N_p(H_1,H_2) \subset \Pi_2(H_1,H_2) = \mathcal{S}_2(H_1,H_2).$$

Clearly if $1 \le p < 2$, $N_p(H_1,H_2) \subset \Pi_2(H_1,H_2)$ so (i)⇒(ii).

(ii)⇒(iii) is trivial.

(iii)⇒(i). Let $T \in N(X)$. Again from the diagram $\delta_1 K_1 \in N_2(X,\ell_2)$.

But $N_2 \subset N_p$ for $p > 2$ so by (iii) $\delta_1 K_1 \in S_2(X,\ell_2)$. Now $\delta_2^* \in N_2(\ell_\infty,\ell_2)$ and so, as above, $\delta_2^* K_2^* \in S_2(X^*,\ell_2)$. But, Hutton [Hu] has shown that $\alpha_n(T) = \alpha_n(T^*)$ for compact T and so $K_2\delta_2 \in S_2(\ell_2,X)$. Thus $T \in S_2 \cdot S_2(X)$ and so has absolutely converging eigenvalues. Observe that reflexivity, used in [R], is not needed.

6. REMARKS.

This concludes the applications of the eigenvalue theorem. Of course there are many more results scattered through the literature which can be proved by this strategy of diagram chasing. This does not necessarily detract from these results nor the theorems discussed here. Most of these results were obtained when we had the "complemented" subspace problem" not the "complemented" subspace theorem". And remember, the eigenvalue theorem has its proof hidden in the complemented subspace theorem. Really, what we have shown is that, unknown to the various authors, at the time, the eigenvalue behavior of nuclear operators on the given space unifies essentially all of the operator characterizations of Hilbert space.

BIBLIOGRAPHY

[C] Cohen, J. (1970). A characterization of inner-product spaces using absolutely 2-summing operators. Studia Math. 38, 271-276.

[D1] Dvoretzky, A. (1961). Some results on convex bodies and Banach spaces. Proc. Int. Symposium on Linear Spaces, Jerusalem 123-160.

[D2] Dvoretzky, A. (1966). A characterization of Banach spaces isomorphic to inner product spaces. Proc. Colloquium Convexity, Univ. of Copenhagen 61-66.

[GLR] Gordon, Y.; Lewis, D.R. and Retherford, J.R. (1973). Banach Ideals of operators with applications. J. Funct. Anal. 14, 85-129.

[G1] Grothendieck, A.(1955). Produits tensoriels et espaces nucléaires Memoir of the Am. Math. Soc. 16.

[G2] Grothendieck, A. (1956). Résumé de la théorie métrique des produits tensoriels topologiques. Bol. Soc. Math. Sao Paulo 8, 1-79.

[G3] Grothendieck, A. (1951). Sur le notion de produit tensoriel topologique d'espaces vectoriels topologique, et une class remarquable d'espaces vectoriels liées a cette notion. C.R. Acad. Sci. Pris 233 1556-1558.

[H] Holub, J.R. (1972). A characterization of subspaces of $L_p(\mu)$. Studia Math. 42, 265-270.

[Hu] Hutton, C.V. (1974). On the approximation numbers of an operator and its adjoint. Math. Ann. 210, 277-280.

[JKMR] Johnson, W.B.; König, H.; Maurey,B. and Retherford, J.R. (1979). Eigenvalues of p-summing and ℓ_p-type operators in Banach spaces. J. Funct. Anal. 32, 380-400.

[K] Kwapién, S. (1970). A linear topological characterization of inner product spaces. Studia Math. 38, 277-278.

[LP] Lindenstrauss, J. and Tzafriri, L. (1971) On the complemented subspaces problem. Israel J. Math. 9 , 263-269.

[MR] Morrell, J.S. and Retherford, J.R. (1972). p-trivial Banach spaces. Studia Math. 48, 1-25.

[P] Pelczynski, A. (1967). A characterization of Hilbert-Schmidt operators. Studia Math. 28 355-360.

[Pi1] Pietsch, A. (1963). Absolut p-summierende Abbildungen in lokal konvexen Räumen. Math. Nachr. 27, 77-103.

[Pi2] Pietsch, A. (1967). Absolut p-summierende Abbildungen in normierten Räumen. Studia Math. 28, 333-353.

[Pi3] Pietsch, A. (1969). Hilbert-Schmidt Abbildungen in Banach Räumen. Math. Nachr., 37, 237-245.

[Pi4] Pietsch, A. (1970). Ideale von S_p-Operatoren in Banachräumen. Studia Math. 38, 59-69.

[Pi5] Pietsch, A. (1978). Operator Ideals. VEB Deutscher Verlag des Wissenschaften, Berlin.

[Pl] Plans, A. (1959). Zerlegung von Folgen im Hilbertraum in Heterogonal Systeme. Archiv der Mathematik, T.X. , 304-306.

[P12] Plans, A. (1961). Propiedades angulares de la convergencia en el espacio de Hilbert. Rev. Mat. Hisp.-Amer., T. XXI , 100-109.

[P13] Plans, A. (1959). Propiedades angulares de los sistemas heterogonales. Rev. Acad. Ciencias de Zaragoza T. XV, 2.

[P14] Plans, A. (1961). Resultados acerca de una generalización de la semejanza en el espacio de Hilbert. Coll. Math. XIII, 241-258.

[P15] Plans, A. (1963). Los operadores acotados en relación con los sistemas asintóticamente ortogonales. Coll. Math. XV, 105-110.

[PB] Plans, A. and Burillo, P. (1975) Operadores Nucleares y Sumabilidad en bases ortonormales. Facultad de Ciencias, Zaragoza.

[R] Rosenberger, B. (1975). Approximationszahlen, p-nucleare Operatoren und Hilberträum-charakterisierungen. Math. Ann. 213, 211-221.

[Ru] Ruston, A.F. (1951). On the Fredholm theory of integral equations for operators belonging to the trace class of a general Banach space. Proc. London Math. Soc. (2), 53, 109-124.

[S] Saphar, P. (1970). Produits tensoriels d'espaces de Banach et classes d'applications lineaires. Studia Math. 38, 71-100.

[W] Weyl, H. (1949). Inequalities between the two kinds of eigenvalues of a linear transformation. Proc. Nat. Acad. Sci. USA 35, 408-411.

PARTICULAR M-BASIC SEQUENCES IN BANACH SPACES.

Paolo Terenzi
Dipart. di Matematica del Pol. di Milano, Milano, Italy.

Dedicated to Antonio Plans on his 65th birthday.

Abstract. We are concerned with the existence of M-bases with weaker properties than the strong M-basis, moreover, with the possibility of decomposing an M-basis in a finite number of subsequences with more regular properties.

INTRODUCTION.

In what follows B is a Banach space; for the definitions of biorthogonal system, minimal, uniformly minimal, M-basis, norming M-basis, strong M-basis, basis with brackets and basis, see [1] or [2] and [3].

Let (x_n) be a sequence with $B = [x_n]$ ($=\overline{\text{span}}\ (x_n)$); we shall consider the representation of the elements of B by means of the elements of (x_n).

We suppose (x_n) minimal, with (x_n, f_n) biorthogonal and $\|x_n\|=1$ for every n. By [4] there exists an increasing sequence $(q_m)_{m=0}^{\infty}$ of integers, with $q_0=0$, such that for every x of B there exists (a_n) of numbers for which

$$(1)\qquad x = \lim_{p\ \infty} \left[\sum_{m=1}^{p} \left[\sum_{n=q_{m-1}+1}^{q_m} f_n(x)x_n \right] + \sum_{n=q_p+1}^{q_{p+1}} a_n x_n \right]$$

where (q_m) does not depend on x.

We shall say that x of B is

(i) <u>regular</u> <u>on</u> (x_n) if $x \in \left[\sum_{n=1}^{m} f_n(x)\, x_n \right]_{m=1}^{\infty}$;

(ii) <u>bounded</u> <u>on</u> (x_n) if it is not $\lim_{m\,\infty} \left\| \sum_{n=1}^{m} f_n(x)x_n \right\| = +\infty$.

A panorama of the representation of B by means of (x_n) is as follows (the definitions of "middle M-basis" and "quasi basis with brackets" are new):

a) (x_n) uniformly minimal $\Longleftrightarrow$ $\lim_{n\,\infty} f_n(x)=0$ for every x of B;

b) (x_n) M-basis $\Longleftrightarrow$ $\sum_{n=1}^{\infty} |f_n(x)| > 0$ for every $x \neq 0$ of B;

c) (x_n) norming M-basis $\Longleftrightarrow$ for every x of

$\|x\| \le K \sup \{ |f(x)|; f\in \text{span } (f_n)\ \|f\|=1\}$, where $K(\ge 1)$ is fixed;

d) (x_n) middle M-basis $\Longleftrightarrow$ every bounded (on (x_n)) element of B is regular on (x_n);

e) (x_n) strong M-basis $\Longleftrightarrow$ every element of B is regular on (x_n);

f) (x_n) quasi-basis with brackets $\Longleftrightarrow$ for every x of B there exist a permutation $(\tilde{n})$ of (n) and an increasing sequence $(p_m)_{m=0}^{\infty}$ of integers, with $p_0 = 0$, such that $x = \sum_{m=0}^{\infty} \left(\sum_{\tilde{n}=p_{m-1}+1}^{p_m} f_{\tilde{n}}(x)\, x_{\tilde{n}} \right)$;

g) (x_n) basis with brackets $\Longleftrightarrow$ $x = \sum_{m=0}^{\infty} \left(\sum_{n=q_{m-1}+1}^{q_m} f_n(x) \right)$ for every x of B, where (q_m) is the sequence of (1);

h) (x_n) basis $\Longleftrightarrow$ $x = \sum_{n=0}^{\infty} f_n(x)x_n$ for every x of B.

We know that every separable B has an uniformly minimal norming M-basis, moreover we know that a separable B in general does not have a basis with brackets; the existence of e) or f) is an open question.

In §1 we consider the existence of the middle basis and of other intermediate M-bases; in §2 we consider the basic subsequences of an M-basis and the possible decomposition in a finite number of subsequences with more regular properties; in §3 there are two open problems.

1. MIDDLE M-BASES AND QUASI-STRONG M-BASES.

Firstly we prove the existence of the middle M-basis in the reflexive spaces.

Theorem 1. *If the dual is separable* B *has a middle M-basis.*

Proof. Let (x_n) be an M-basis of B. Let (g_n) be dense in the unit sphere of B^*.

Let q_1 be the first integer such that $g_1(x_{q_1}) \neq 0$, we set

$$y_n = x_n \text{ for } 1 \le n \le q_1 \quad , \quad x_{1n} = x_n - \frac{g_1(x_n)}{g_1(x_{q_1})} x_{q_1} \quad \text{for } n > q_1;$$

consider g_2, if $(x_{1n})_{n>q_1} \subset g_{2\perp}$ we consider g_3, otherwise there exists $q_2 > q_1$ such that $g_2(x_{1q_2}) \neq 0$, then we set

$$y_n = x_{1n} \quad \text{for } q_1+1 \le n \le q_2 \; ,$$

$$x_{2n} = x_{1n} - \frac{g_2(x_{1n})}{g_2(x_{1q_2})} x_{1q_2} \qquad \text{for } n > q_2.$$

So proceeding we get (y_n) and a not decreasing sequence (ℓ_n) of integers, so that for every m

$$\text{span}\,(y_n)_{n=1}^{m} = \text{span}\,(x_n)_{n=1}^{m} \quad , \quad (y_n)_{n>\ell_m} \subset g_{m\perp}. \tag{2}$$

Since (y_n) is obviously minimal it follows that (y_n) is M-basis of B, indeed, by (2) if $y \in \bigcap_{m=1}^{\infty} [y_n]_{n\geq m}$ we have $g_n(y) = 0$ for every n, hence y=0. If (y_n) is not strong M-basic, by [3] (p. 243, prop. 8.11, 3) we have that there exist $\bar{x}$ of B and $(n)=(n_k)\cup(n'_k)$, $(n_k)\cap(n'_k) = \emptyset$, so that setting $Y'=[y_{n'_k}]$,

$$\bar{x} + Y' \in \bigcap_{m=1}^{\infty} [y_{n_k} + Y']_{k\geq m}, \quad \|\bar{x} + Y'\| = 1.$$

Then there exist $\bar{F}$ of $(B/Y')^*$ and $\bar{f}$ of B^* so that

$$\bar{F}(\bar{x}+Y') = \|\bar{F}\| = 1 = \|\bar{f}\| = \bar{f}(\bar{x}), \text{ where } \bar{f}(x) = \bar{F}(x+Y') \ \forall x \in B. \tag{3}$$

Let (g_{i_p}) be a subsequence of (g_n) such that

$$\|g_{i_p} - \bar{f}\| < 1/2^p \text{ for every p.} \tag{4}$$

We shall prove that $\bar{x}$ is not bounded on (y_n), which will prove the thesis. Indeed suppose $\bar{x}$ bounded, there exist two increasing sequences (r_m) and (t_m) of integers such that, for every m,

$$\left\| \sum_{k=1}^{r_m} f_k(\bar{x})\, y_k \right\| = \left\| \sum_{k=1}^{t_m} f_{n'_k}(\bar{x})\, y_{n'_k} \right\| \leq a < +\infty. \tag{5}$$

For every integer $p \geq 1$ there exists

$$\tilde{x}_p = \sum_{k=1}^{t_{m(p)}} f_{n'_k}(\bar{x})\, y_{n'_k} + \sum_{k=s_{p-1}+1}^{s_p} a_{pk} y_k, \text{ with} \tag{6}$$

$$s_{p-1} = n'_{t_{m(p)}} > \ell_{i_p}, \quad \|\tilde{x}_p - \bar{x}\| < \frac{1}{2^p}.$$

By (2), (6), (3), (4) and (5) it follows that

$$\lim_{p\to\infty}\left|g_{i_p}(\tilde{x}_p)\right| = \lim_{p\to\infty}\left|g_{i_p}\left(\sum_{k=1}^{t_{m(p)}} f_{n'_k}(\bar{x})\, y_{n'_k}\right)\right| =$$

$$= \lim_{p\to\infty}\left|\left(g_{i_p}-\bar{f}\right)\left[\sum_{k=1}^{t_{m(p)}} f_{n'_k}(\bar{x})\, y_{n'_k}\right]\right| \le$$

$$\le \lim_{p\to\infty}\|g_{i_p} - \bar{f}\|\cdot\left\|\left(\sum_{k=1}^{t_{m(p)}} f_{n'_k}(\bar{x})\, y_{n'_k}\right)\right\| \le \lim_{p\to\infty}\frac{a}{2^p} = 0;$$

on the other hand, by (3), (6) and (4),

$$\lim_{p\to\infty}|g_{i_p}(\tilde{x}_p)| = \lim_{p\to\infty}|\bar{f}(\bar{x}) + \bar{f}(\tilde{x}_p-\bar{x}) + (g_{i_p} - \bar{f})(\bar{x}+(\tilde{x}_p-\bar{x}))|$$

$$\ge 1 - \lim_{p\to\infty}|\bar{f}(\tilde{x}_p-\bar{x}) + (g_{i_p} - \bar{f})(\bar{x} + (\tilde{x}_p-\bar{x}))|$$

$$\ge 1 - \lim_{p\to\infty}\left[\|\tilde{x}_p-\bar{x}\| + \|g_{i_p} - \bar{f}\|\left(\|\bar{x}\| + \|\tilde{x}_p-\bar{x}\|\right)\right]$$

$$\ge 1 - \lim_{p\to\infty}\left(\frac{1}{2^p} + \frac{1}{2^p}\left(\|\bar{x}\| + \frac{1}{2^p}\right)\right) = 1;$$

therefore x cannot be bounded on (y_n), which completes the proof of th. I.

Let us consider another intermediate sequence between the M-basis and the strong M-basis. We know that

$$(x_n) \text{ is strong M- basic} \iff (x_{n_k}+[x_{n'_i}]_{i=1}^{\infty})_{k=1}^{\infty} \text{ is M-basic for}$$

$$\text{every } (n) = (n_k)_{k=1}^{\infty} \cup (n'_k)_{k=1}^{\infty}, \quad (n_k)\cap(n'_k)=\emptyset.$$

Then we say that (x_n) is <u>numerably strong M-basic</u> if there exists a numerable sequence of partitions of (n)

$$N_i = (n_{ik})_{k=1}^{\infty} \cup (n'_{ik})_{k=1}^{\infty} \ , \ (n_{ik}) \cap (n'_{ik}) = \emptyset \ , \ i=1,2,\ldots;$$

such that

$$(x_{n_{ik}} + [x_{n'_{ij}}]_{j=1}^{\infty})_{k=1}^{\infty} \quad \text{is M-basic for } i=1,2,\ldots.$$

Next theorem shows the existence of this intermediate type of M-bases.

Theorem II. *Let* (x_n) *be a sequence of* B *with* $[x_n]$ *of infinite dimension and let* (N_i) *be a numerable sequence of partitions of* (n)

$$N_i = (n_{ik})_{k=1}^{\infty} \cup (n'_{ik})_{k=1}^{\infty} \ , \ (n_{ik}) \cap (n'_{ik}) = \emptyset \ , \ i=1,2,\ldots$$

Then there exists (y_n) of $[x_n]$ *such that*

(i) (y_n) *is norming M-basis of* $[x_n]$ *with* $(y_n)_{n\geq m} \subset \text{span}(x_n)_{n\geq q_m}$ *for every* m, *where* (q_m) *is a non decreasing sequence of integers with* $q_m \longrightarrow +\infty$;

(ii) (y_n) *is numerably strong M-basic along the sequence* (N_i) *of partitions of* (n).

Proof. By th. I of [5] we can suppose (x_n) norming M-basic.
Consider i=1.
We have

$$(x_n) = x_1 \cup (x_{n_{1k}})_{k\geq t_1} \cup (x_{n'_{1k}})_{k\geq t'_1}.$$

Set

$$y_1 = x_1 \ , \ x_{1n'_{1k}} = x_{n'_{1k}} \quad \text{for } k \geq t'_1,$$

$$X'_1 = [x_{n'_{1k}}]_{k\geq t'_1} \ , \ \bar{X}_1 = \bigcap_{m=t_1}^{\infty} [x_{n_{1k}} + X'_1]_{k\geq m} \ ;$$

suppose that

$$(\bar{x}_{1n} + X'_1)_{n=1}^{\infty} \quad \text{is an M-basis of } \bar{X}_1 \ , \text{ with } (\bar{x}_{1n} + X'_1, \bar{F}_{1n})$$

$$(\bar f_{1n}) \subset B^* \quad \text{so that } \bar f_{1n}(x) = \bar F_{1n}(x+X'_1) \text{ for every } x \text{ of } B \text{ and for every } n.$$

biorthogonal,

Consider $\bar f_{11}$.

There exists an increasing sequence $(p(1,1,k))$ of integers such that $\bar f_{11}(x_{n_{1p(1,1,k)}}) \neq 0$ for every k (otherwise it would be $\bar X_{1,1} \subset \bar f_{11_\perp}$), then we set

$$x_{1n_{1k}} = x_{n_{1k}} \quad \text{for } t_1 \leq k \leq p(1,1,1);$$

moreover, for every k and for $p(1,1,k)+1 \leq i \leq p(1,1,k+1)$,

$$x'_{1n_{1i}} = x_{n_{1i}} - \frac{\bar f_{11}(x_{n_{1i}})}{\bar f_{11}(x_{n_{1p(1,1,k)}})} x_{n_{1p(1,1,k)}}.$$

Then $\bar f_{11}(x'_{1n_{1k}}) = 0$ for $k > p(1,1,1)$; that is

$$\bigcap_{m=p(1,1,1)+1}^{\infty} \left[x'_{1n_{1k}} + X'_1\right]_{k\geq m} \subseteq \left[\bar x_{1n} + X'_1\right]_{n\geq 2} \subset \bar F_{11_\perp}.$$

Consider $\bar f_{12}$. If $\bigcap_{m\geq p(1,1,1)} \left[x'_{1n_{1k}} + X'_1\right]_{k\geq m} \not\subset \bar F_{12_\perp}$ there exists an increasing sequence $(p(1,2,k))_{k=1}^{\infty}$ of integers such that $\bar f_{12}(x'_{1n_{1p(1,2,k)}}) \neq 0$ for every k, hence we set

$$x_{1n_{1k}} = x'_{1n_{1k}} \quad \text{for } p(1,1,1)+1 \leq k \leq p(1,2,1);$$

moreover, for every k and for $p(1,2,k)+1 \leq i \leq p(1,2,k+1)$,

$$x''_{1n_{1i}} = x'_{1n_{1i}} - \frac{\bar{f}_{12}(x'_{1n_{1i}})}{\bar{f}_{12}(x'_{1n_{1p(1,2,k)}})} x'_{1n_{1p(1,2,k)}} .$$

So proceeding we get $y_1 \cup (x_{1n})_{n>1}$ such that

$$\text{span}\,(x_{1n})_{n>1} = \text{span}\,(x_n)_{n>1} ;$$

$(x_{1n})_{n\geq m} \subset \text{span}\,(x_n)_{n\geq q_{1m}}$ for every m, where (q_{1m}) is a non decreasing sequence of integers with $q_{1m} \longrightarrow +\infty$;

$(x_{1n_{1k}} + [x_{1n'_{1j}}]_{j=t'_1}^{\infty})_{k=t_1}^{\infty}$ is M-basic.

Consider i=2.
We have

$$(x_{1n})_{n>2} = (x_{1n_{1k}})_{k\geq s_1} \cup (x_{1n'_{1k}})_{k\geq s'_1}$$
$$= (x_{1n_{2k}})_{k\geq s_2} \cup (x_{1n'_{2k}})_{k\geq s'_2}$$

Set $y_2 = x_{1,2}$; $x_{2n'_{2k}} = x_{1n'_{2k}}$ for $k \geq s'_2$.

Suppose that

$$X'_2 = [x_{1n'_{2k}}]_{k\geq s'_2} , \quad \bar{X}_2 = \bigcap_{m=s_2}^{\infty} \left[x_{1n_{2k}} + [x_{1n'_{2j}}]_{j=s'_2}^{\infty} \right]_{k\geq m}$$

$(\bar{x}_{2n} + X'_2)_{n=1}^{\infty}$ M-basis of $\bar{X}_2$ with $(\bar{x}_{2n} + X'_2, \bar{F}_{2n})_{n=1}^{\infty}$ biorthogonal,

$(\bar{f}_{2n})_{n=1}^{\infty} \subset B^*$ so that $\bar{f}_{2n}(x) = \bar{F}_{2n}(x + X'_2)$ for every x of B and for every n.

Consider $\bar{f}_{21}$.
We have that

$$(n_{2k})_{k\geq s_2} = (n^{(1)}_{2k})_{k\geq s_2^{(1)}} \cup (n^{(2)}_{2k})_{k\geq s_2^{(2)}} \text{ with}$$

$$(7) \qquad (n^{(1)}_{2k})_{k\geq s_2^{(1)}} = (n_{2k})_{k\geq s_2} \cap (n_{1k})_{k\geq s_1},$$

$$(n^{(2)}_{2k})_{k\geq s_2^{(2)}} = (n_{2k})_{k\geq s_2} \cap (n'_{1k})_{k\geq s'_1}.$$

Consider two increasing sequences of integers $(p^{(i)}(2,1,k))_{k=1}^{\infty}$ $(i=1,2)$ so that

$$\bar{f}_{21}(x_{1n^{(i)}_{2p^{(i)}(2,1,k)}}) \neq 0 \text{ for every } k \text{ and for } i=1,2$$

(maybe that one of these two sequences of integers does not exist, for example for $i=1$, then we shall set $x_{2n_{2k}}(1) = x_{1n_{2k}}(1)$ for $k\geq s_2^{(1)}$ and we shall consider only $i=2$). Now we set, for $i=1,2$,

$$x_{2n^{(i)}_{2k}} = x_{1n^{(i)}_{2k}} \quad \text{for} \quad s_2^{(i)} \leq k \leq p^{(i)}(2,1,1);$$

moreover, for every k and for $p^{(i)}(2,1,k) + 1 \leq j \leq p^{(i)}(2,1,k+1)$,

$$(8) \qquad x'_{2n^{(i)}_{2j}} = x_{1n^{(i)}_{2j}} - \frac{\bar{f}_{21}(x_{1n^{(i)}_{2j}})}{\bar{f}_{21}(x_{1n^{(i)}_{2p(2,1,k)}})} x_{1n^{(i)}_{2p(2,1,k)}}.$$

Then $\bar{f}_{21}(x'_{2n_{2k}}) = 0$ for large k; moreover $(x'_{2n_{1k}} + [x'_{2n'_{1j}}]^{\infty}_{j=s'_1})_{k\geq s_1}$ is still M-basic because by (7) and (8) we have that

$$[x'_{2n'_{1j}}]^{\infty}_{j=s'_1} = [x_{1n'_{1j}}]^{\infty}_{j=s'_1};$$

$(x'_{2n_{1k}})_{k\geq m} \subset \operatorname{span}(x_{1n_{1k}})_{k\geq q(1,m)}$ for every m, where $(q(1,m))$ is a non decreasing sequence of integers with $q(1,m) \longrightarrow +\infty$.

In the same way we consider $\bar{f}_{22}$ and so on.

So proceeding we get $(y_1, y_2) \cup (x_{2n})_{n>2}$ such that

$$\mathrm{span}\,(x_{2n})_{n>2} = \mathrm{span}\,(x_{1n})_{n>2} ;$$

$(x_{2n})_{n\geq m} \subset \mathrm{span}\,(x_{1n})_{n\geq q_{2m}}$ for every $m > 2$, where (q_{2m}) is a non decreasing sequence of integers with $q_{2m} \longrightarrow +\infty$;

$(x_{2n_{1k}} + [x_{2n'_{1j}}]^{\infty}_{j=s'_1})^{\infty}_{k=s_1}$ is M-basic;

$(x_{2n_{2k}} + [x_{2n'_{2j}}]^{\infty}_{j=s'_2})$ is M-basic.

Consider i=3.
We set $y_3 = x_{2,3}$; then we have

$$(x_{2n})_{n>3} = (x_{2n_{1k}})_{k\geq r_1} \cup (x_{2n'_{1k}})_{k\geq r'_1} =$$
$$= (x_{2n_{2k}})_{k\geq r_2} \cup (x_{2n'_{2k}})_{k\geq r'_2} =$$
$$= (x_{2n_{3k}})_{k\geq r_3} \cup (x_{2n'_{3k}})_{k\geq r'_3} .$$

Set $x_{3n'_{3k}} = x_{2n'_{3k}}$ for $k \geq r'_3$. Suppose that

$$X'_3 = [x_{2n'_{3k}}]_{k\geq r'_3} .\ \ \bar{X}_3 = \bigcap_{m=r_3}^{\infty} [x_{2n_{3k}} + [x_{2n'_{3j}}]^{\infty}_{j=r'_3}]_{k\geq m} ,$$

$(\bar{x}_{3m} + X'_3)^{\infty}_{n=1}$ M-basis of $\bar{X}_3$ with $(\bar{x}_{3n} + X'_3, \bar{F}_{3n})$ biorthogonal,

$(\bar{f}_{3n})^{\infty}_{n=1} \subset B^*$ so that $\bar{f}_{3n}(x) = \bar{F}_{3n}(x+X'_3)$ for every x of B and for every n.

Consider $\bar{f}_{31}$.
We have that

$$(n_{3k})_{k\geq r_3} = \bigcup_{i=1}^{4} (n^{(i)}_{3k})_{k\geq r^{(i)}_3} \quad \text{where}$$

$$(n^{(1)}_{3k})_{k\geq r^{(1)}_3} = (n_{3k})_{k\geq r_3} \cap (n'_{1k})_{k\geq r'_1} \cap (n'_{2k})_{k\geq r'_2},$$

$$(n^{(2)}_{3k})_{k\geq r^{(2)}_3} = (n_{3k})_{k\geq r_3} \cap (n'_{1k})_{k\geq r'_1} \cap (n_{2k})_{k\geq r_2},$$

$$(n^{(3)}_{3k})_{k\geq r^{(3)}_3} = (n_{3k})_{k\geq r_3} \cap (n_{1k})_{k\geq r_1} \cap (n'_{2k})_{k\geq r'_2},$$

$$(n^{(4)}_{3k})_{k\geq r^{(4)}_3} = (n_{3k})_{k\geq r_3} \cap (n_{1k})_{k\geq r_1} \cap (n_{2k})_{k\geq r_2},$$

We consider again four sequences of integers $(p^{(i)}(3,1,k))^{\infty}_{k=1}$ $(i=1,2,3,4)$ so that

$$\bar{f}_{31}(x_{2n^{(i)}_{3p^{(i)}(3,1,k)}}) \neq 0 \text{ for every } k \text{ and for } i=1,\dots,k\ .$$

If some of these sequences of integers do not exist, then we shall proceed as for $(p^{(i)}(2,1,k))$. It is now sufficient to set , for $i=1,\dots,4$,

$$x_{3n^{(i)}_{3k}} = x_{2n^{(i)}_{3k}} \quad \text{for} \quad r^{(i)}_3 \leq k \leq p^{(i)}(3,1,1)\ ;$$

moreover, for every k and for $(p^{(i)}(3,1,k)+1 \leq j \leq p^{(i)}(3,1,k+1)$,

$$x'_{3n^{(i)}_{3j}} = x_{2n^{(i)}_{3j}} - \frac{\bar{f}_{31}(x_{2n^{(i)}_{3j}})}{\bar{f}_{31}(x_{2n^{(i)}_{3p^{(i)}(3,1,k)}})} x_{2n^{(i)}_{3p^{(i)}(3,1,k)}}\ .$$

Now the procedure is clear, for $(\bar{f}_{3n})_{n>1}$ and for $i>3$.
So proceeding we get (y_n) as in the thesis (indeed, since (x_n) is norming and since $(y_n)_{n\geq m} \subset \text{span}\,(x_n)_{n\geq q_m}$ with $q_m \longrightarrow +\infty$, (y_n) is norming too). This completes the proof of th. II.

We shall say that an M-basic sequence (x_n) is <u>quasi-strong M-basic</u> if (x_n) has the possibility to become strong M-basic by means of the procedure of Th. II; that is if, after removing the not regular points along a numerable sequence of partitions of (n), the sequence becomes strong M-basic.

About the stability we point out that a middle (numerably strong) (quasi-strong) (strong) M-basic sequence keeps its properties for sufficiently "near" sequences: it is sufficient to prove this for the strong M-bases.

Proposition. *If* (x_n) *is strong M-basic in* B, *there exists* (ε_n) *of positive numbers such that every* (y_n) *of* B *with* $\|y_n - x_n\| < \varepsilon_n$ *for every* n *is strong M-basic.*

Proof. If (x_n, f_n) is biorthogonal suppose that

$$\|y_n - x_n\| < \varepsilon_n = 1/(2^{n+1}\|f_n\|) \text{ for every } n.$$

Following the techniques of the proof of the Krein-Milman-Rutman theorem (see [2] p. 84-99) it immediately follows, for every $(a_n)_{n=1}^m$, of numbers,

$$(9) \qquad \frac{1}{2}\left\|\sum_{n=1}^m a_n x_n\right\| \le \left\|\sum_{n=1}^m a_n y_n\right\| \le \frac{3}{2}\left\|\sum_{n=1}^m a_n x_n\right\|$$

If (y_n) is not strong M-basic there exist $(n) = (n_k) \cup (n'_k)$, $(n_k) \cap (n'_k) = \emptyset$ and $\bar{y}$ of B so that

$$(10) \qquad \left\|\bar{y} + [y_{n'_k}]\right\| = 1 \; ; \; \left\|\bar{y} - \left(\sum_{k=1}^{r_p} a_{pk}\, y_{n'_k} + \sum_{k=s_{p-1}+1}^{s_p} b_k y_{n_k}\right)\right\| < \frac{1}{2^p}$$

for every p where $(s_p)_{p=0}^{\infty}$ is an increasing sequence of integers with $s_0=0$. Set for every p

$$\tilde{x}_p = \sum_{k=1}^{r_p} a_{pk}\, x_{n'_k} + \sum_{k=s_{p-1}+1}^{s_p} b_k\, x_{n_k}$$

by (9) and (10) there exists $\bar{x}$ of $[x_n]$ so that

$$\lim_{p\to\infty} \tilde{x}_p = \bar{x} \ , \quad \frac{2}{3} \le \| \bar{x} + [x_{n'_k}] \| \le 2 \ ;$$

hence (x_n) would not be strong M-basic, which completes the proof of the proposition.

2. BASIC SUBSEQUENCES AND DECOMPOSITION IN A FINITE NUMBER OF MORE REGULAR SUBSEQUENCES.

Firstly we consider the "extraction" of basic subsequences from a sequence. It is well known that every M-basic sequence has an infinite basic subsequence. The presence of these basic subsequences is very "visible" in the norming and uniformly minimal case, as follows:

Theorem III. *If* (x_n) *is norming and uniformly minimal there exists an increasing sequence* $(q_m)_{m=0}^{\infty}$ *of integers with* $q_0=0$ *such that every* $(x_{n_m})_{m=1}^{\infty}$ *with* $q_{m-1}+1 \le n_m \le q_m$ *for every m is basic.*

Proof. Since (x_n) is norming we can choose the sequence (q_m) of (1) so that, for every m.

$$\text{dist}\ (x, [x_n]_{n>q_{m+1}}) \ge \|x\| \ (1 - \frac{1}{2^m}) \text{ for every } x \in [x_n]_{n=1}^{q_m}$$

Consider (x_{n_m}) with $q_{m-1}+1 \le n_m \le q_m$ for every m.

Let $\bar{x} \in [x_{n_m}]$ with $\|\bar{x}\| = 1$, by (1) and since (x_n) is uniformly minimal there exists a subsequence (p(i)) of (i) such that, for every $p \ge p(i)$ and for every i,

$$\left\| \bar{x} - \left(\sum_{m=1}^{p} f_{n_m}(\bar{x})\, x_{n_m} + \sum_{m=q_p+1}^{q_{p+1}} \bar{a}_m x_m \right) \right\| < \frac{1}{2^i} \; ; \; \left| f_{n_{p+1}}(\bar{x}) \right| < \frac{1}{2^i} \; ;$$

$$\left\| \bar{x} - \sum_{m=1}^{p} f_{n_m}(\bar{x})\, x_{n_m} + x \right\| > \left\| \bar{x} - \sum_{m=1}^{p} f_{n_m}(\bar{x})\, x_{n_m} \right\| \left(1 - \frac{1}{2^i}\right) \text{ for}$$

$$\text{every } x \in [x_n]_{n>q_{p+1}} .$$

It follows that

$$\left\| \bar{x} - \sum_{m=1}^{p} f_{n_m}(\bar{x})\, x_{n_m} \right\| < \left\| \bar{x} - \sum_{m=1}^{p} f_{n_m}(\bar{x})\, x_{n_m} - \sum_{m=q_{p+1}+1}^{q_{p+2}} \bar{a}_m x_m \right\| \frac{1}{1 - \frac{1}{2^i}}$$

$$< \left(\left\| \bar{x} - \sum_{m=1}^{p+1} f_{n_m}(\bar{x})\, x_{n_m} - \sum_{m=q_{p+1}+1}^{q_{p+2}} \bar{a}_m x_m \right\| + \frac{1}{2^i} \right) \frac{1}{1 - \frac{1}{2^i}} <$$

$$< \frac{1}{2^{i-1}} \; \frac{1}{1 - \frac{1}{2^i}}$$

for every i, that is $\bar{x} = \sum_{m=1}^{\infty} f_{n_m}(\bar{x})\, x_{n_m}$, which completes the proof.

We pass to consider the possible decomposition of an M-basis in a finite number of more regular subsequences.

It is known that, if (x_n) is norming M-basic, the sequence (q_m) of (1) can be selected so that

$$\bigcup_{m=1}^{\infty} (x_n)_{n=q_{2m-1}+1}^{q_{2m}} \quad \text{and} \quad \bigcup_{m=0}^{\infty} (x_n)_{n=q_{2m}+1}^{q_{2m+1}}$$

are basic with brackets.

In the example of [4] we proved that in general an M-basic

sequence (x_n) cannot be decomposed in a finite number of basic with brackets subsequences (also if (x_n) is quasi-basic with brackets).

In a recent seminar at Zaragoza the following questions were raised:

Question 1. Is it possible to decompose every M-basic sequence in 2 (or in a finite number of) strong M-basic subsequences?

Question 2. Does there exist in every M-basic sequence a strong M-basic (or basic) subsequence which is "maximal" (that is, by adding infinite elements the sequence loses its properties)?

Next three examples give negative answers to question 1.

In what follows $(x_n) = \bigcup_{m=1}^{\infty} (x_{mn})_{n=1}^{\infty}$ is algebraic basis of a linear space L; (y_n) is M-basis of a Banach space Y, with $Y \cap L = \emptyset$.

EXAMPLE 1. Suppose that (y_n) is not strong M-basic and set for every $\bigcup_{m=1}^{p} (a_{mk})_{k=1}^{p}$ of numbers

$$(11) \qquad \left\| \sum_{m=1}^{p} \left(\sum_{k=1}^{p} a_{mk} x_{mk} \right) \right\| = \left\| \sum_{m=1}^{p} \left\| \sum_{k=1}^{m-1} (a_{mk} x_{mk} + a_{km} x_{km}) + a_{mm} x_{mm} \right\| = \right.$$

$$= \left\| \sum_{m=1}^{p} \left(\left\| \sum_{k=1}^{m-1} (|a_{mk}| + |a_{km}|) + |a_{mm}| \right) y_m \right\| \right.$$

This defines a norm on span (x_n), moreover (x_n) is minimal since, beeing (y_n) minimal, there exists (δ_n) of positive numbers such that, for every $(a_n)_{n=1}^{m}$ of numbers,

$$\left\| \sum_{n=1}^{m} a_n y_n \right\| \geq \sum_{n=1}^{m} |a_n| \, \delta_n \ ;$$

hence for every $\bigcup_{m=1}^{p} (a_{mk})_{k=1}^{p}$ of numbers by (11) it follows that

$$\left\| \sum_{m=1}^{p} \left(\sum_{k=1}^{p} a_{mk} x_{mk} \right) \right\| \geq \sum_{m=1}^{p} \left(\sum_{k=1}^{p} |a_{mk}| \delta_k \right) \delta_m .$$

Moreover (x_n) is M-basis of $X = [x_n]$, since otherwise there would exist $\bar{x}$ of X, $\|\bar{x}\| = 1$, an increasing sequence $(p(s))$ of integers and a sequence $(\bar{a}_n)$ of numbers so that, for every s

$$\left\| \bar{x} - \sum_{m=p(s)+1}^{p(s+1)} \left(\sum_{k=1}^{m-1} (\bar{a}_{mk} x_{mk} + \bar{a}_{km} x_{km}) + \bar{a}_{mm} x_{mm} \right) \right\| < \frac{1}{2^{s+1}} ,$$

hence for every i

$$\left\| \sum_{m=p(s)+1}^{p(s+1)} \left(\sum_{k=1}^{m-1} (\bar{a}_{mk} x_{mk} + \bar{a}_{km} x_{km}) + \bar{a}_{mm} x_{mm} \right) - \right.$$

$$\left. - \sum_{m=p(s+i)+1}^{p(s+i+1)} \left(\sum_{k=1}^{m-1} (\bar{a}_{mk} x_{mk} + \bar{a}_{km} x_{km}) + \bar{a}_{mm} x_{mm} \right) \right\| < \frac{1}{2^{s}};$$

therefore, setting for $p(s)+1 \leq m \leq p(s+1)$ and for $p(s+i)+1 \leq m \leq p(s+i+1)$

$$b_m = \sum_{k=1}^{m-1} \left(|\bar{a}_{mk}| + |\bar{a}_{km}| \right) + |\bar{a}_{mm}|$$

by (11) it follows that

$$\left\| \sum_{m=p(s)+1}^{p(s+1)} b_m y_m \right\| > 1 - \frac{1}{2^{s+1}} , \quad \left\| \sum_{m=p(s)+1}^{p(s+1)} b_m y_m - \sum_{m=p(s+i)+1}^{p(s+i+1)} b_m y_m \right\| < \frac{1}{2^{s}} ,$$

which is impossible since (y_n) is M-basic.

We claim that (x_n) cannot be decomposed in two strong M-basic subsequences.

We point out by (11), that for every m and n, $(x_{mn})_{n=1}^{\infty}$ and $(x_{mn})_{m=1}^{\infty}$ are not strong M-basic. Suppose $(x_n) = (x_{n_k})_{k=1}^{\infty} \cup (x_{n'_k})_{k=1}^{\infty}$ with $(x_{n'_k})$ strong M-basic; since (y_n) is not strong M-basic, by (11) there exists an infinite subsequence $(m(i))$ of (m) such that

$$\bigcup_{i=1}^{\infty} (x_{m(i)n})_{n=1}^{\infty} \subseteq (x_{n'_k}) \;;$$

therefore $(x_{n'_k})$ cannot be strong M-basic, since every $(x_{m(i)n})_{n=1}^{\infty}$ is not strong M-basic, for every i. This completes the example.

EXAMPLE 2. As in example 1, where now we can suppose (by example 1) that it is not possible to decompose (y_n) in two strong M-basic subsequences. Therefore now, for every m and n, it is not possible to decompose $(x_{mn})_{n=1}^{\infty}$ and $(x_{mn})_{m=1}^{\infty}$ in two strong M-basic subsequences.

Then, proceeding as in example 1, if $(x_n) = (x_{n_k})_{k=1}^{\infty} \cup (x_{n'_k})_{k=1}^{\infty}$ with (x_{n_k}) strong M-basic, it is not possible to decompose $(x_{n'_k})$ in two strong M-basic subsequences. That is, it is not possible to decompose (x_n) in three strong M-basic subsequences. So proceeding for every m there exists an M-basic sequence which cannot be decomposed in m strong M-basic subsequences.

EXAMPLE 3. Suppose that, for every m, $(y_{mn})_{n=1}^{\infty}$ is M-basis of a Banach space Y_m, so that it is not possible to decompose (y_{mn}) in m strong M-basic subsequences, moreover suppose that $\overline{Y_1+Y_2+\ldots+Y_m+\ldots} \cap L = \emptyset$. We set for every $\bigcup_{m=1}^{p} (a_{mn})_{n=1}^{p_m}$ of numbers

$$\left\| \sum_{m=1}^{p} \left(\sum_{n=1}^{p_m} a_{mn} x_{mn} \right) \right\| = \sum_{m=1}^{p} \left\| \sum_{n=1}^{p_m} a_{mn} y_{mn} \right\|$$

Then (x_n) is M-basic of $X = [x_n] = \sum_{m=1}^{\infty} [x_{mn}]_{n=1}^{\infty}$; but it is not possible to decompose (x_n) in a finite number of strong M-basic subsequences

Finally the next example gives negative answer to question 2.

EXAMPLE 4. If (x_n) and (y_n) are as above, set now for every $\bigcup_{m=1}^{p} (a_{mk})_{k=1}^{p_m}$ of numbers

$$(12) \qquad \left\| \sum_{m=1}^{p} \left(\sum_{k=1}^{p_m} a_{mk} x_{mk} \right) \right\| = \left\| \sum_{m=1}^{p} \left(\sum_{k=1}^{p_m} |a_{mk}| \right) y_m \right\|.$$

Set $X = [x_n]$, it is easy to see that (x_n) is an M-basis of X.
Let (x_{n_k}) be a strong M-basis (basic) subsequence of (x_n), with

$$(x_{n_k}) = \bigcup_{m=1}^{\infty} (x_{p(m)q(k)})_{k=1}^{\infty} ;$$

it cannot be $(p(m))=(m)$ otherwise by (12) (x_{n_k}) would not be strong M-basic, since (y_n) is not strong M-basic.

Therefore if $\bar{m} \notin (p(m))$ we have that $(x_{n_k}) \cup (x_{\bar{m}k})_{k=1}^{\infty}$ is still strong M-basic (basic). Hence (x_n) does not have a "maximal" strong M-basic (basic) subsequence.

3. OPEN PROBLEMS.

About the middle M-bases:

Problem 1. Does there exist in every separable Banach space a middle M-basis? This is a weaker problem than the existence of the strong M-basis.

About the quasi-strong M-basis: we don't know an example of an M-basis which is not quasi-strong M-basis, hence.

Problem 2. Is every M-basis a quasi-strong M-basis ?

By th. II a positive answer to problem 2 implies positive answer to existence of the strong M-basis.

REFERENCES.

[1] Lindenstrauss, J. & Tzafriri, L. (1977). Classical Banach spaces I. Sequence spaces. Springer-Verlag 1977.
[2] Singer, I. (1970). Bases in Banach spaces I. Springer-Verlag.
[3] Singer, I. (1981). Bases in Banach spaces II. Springer-Verlag.
[4] Terenzi, P. (1984). Representation of the space spanned by a sequence in a Banach space. Archiv. der Mathematik 43, pp. 448-459.
[5] Terenzi, P. (1987). On the theory of fundamental norming bounded biorthogonal systems in Banach spaces. Transactions of the A.M.S. (299) 2, p. 497-509.

BEHAVIOUR OF SEMI-FREDHOLM OPERATORS ON A HILBERT CUBE

Alvaro A. Rodés Usán.
Departamento de Matemáticas. Universidad de Zaragoza.

A Antonio Plans, mi maestro y amigo.

0. INTRODUCTION

Let **H** be a real separable Hilbert space. Let **B**, $\mathfrak{J}$, $\mathfrak{J}_1$ be the set of bounded linear, compact and nuclear operators on H respectively.

There are several characterizations through absolute summability of distinct types of operators of B. For instance

$A \in \mathfrak{J}_1$ if and only if $\exists(e_n)$ orthonormal basis (o.n.b.) of **H**, $\sum \|Ae_n\| < \infty$ [Ho]

$\sum \|Ae_n\| < \infty$ for every o.n.b. of **H** if and only if $A=0$ [Bu]

$$A \in \mathfrak{J}_1, \ \inf\left\{ \sum \|Ae_n\|;\ (e_n)\ \text{(o.n.b.) of } \mathbf{H} \right\} = \sum \|A\bar{e}_n\| < \infty$$

where $(\bar{e}_n)$ denotes the orthonormal basis of singular vectors of A (i.e eigenvectors of $(A^*A)^{1/2}$). [Go-Ma]

If we change the role of the unit sphere (support set of all the orthonormal basis) by that of the Hilbert cube, we obtain the Macaev's ideal $\mathfrak{J}_\omega$ defined by

$$\mathfrak{J}_\omega = \left\{ A \in \mathfrak{J};\ \sum 1/n\ \|A\bar{e}_n\| < \infty \right\} \text{[Go-K]}$$

Since A is linear, $\sum 1/n\ \|A\bar{e}_n\| = \sum \|A\ 1/n\ \bar{e}_n\|$, so the Macaev's ideal is made up of all the compact operators which give absolute summability over the vectors $(\bar{e}_n/n)$ which are precisely the central points of the faces of the Hilbert cube associated to the o.n.b. of singular vectors

$$Q(\bar{e}_n) = \left\{ \bar{x} \in H;\ \bar{x} = \sum x_n \bar{e}_n,\ |x_n| \le 1/n \right\}$$

In this paper we have started to study the set of operators of **B** which give absolute summability on vector sequences of a Hilbert cube.

The set of o.n. bases can be considered as the set of intersections of the unit sphere of H with distinct complete orthogonal systems of rays (r_n) of **H**, i.e.

$(e_n) = S \cap (r_n)$
(we take only one of the two intersection vectors)

However there are as many Hilbert cubes as ordered o.n. basis. For a fixed Hilbert cube Q, the relative position of a complete orthogonal system of rays (r_n) depends obviously on the set of rays; there are even complete orthogonal systems of rays which intersect Q only in the null vector, and in this case every operator of **B** gives clearly absolute summability on the intersection vectors of $(r_n) \cap Q$. The problem will be non trivial only when

$(r_n) \cap Q \neq 0,$

for ininitely many values of n.

The simplest case (we shall concentrate our attention on it) is when the rays (r_n) intersect Q in the central points of its faces, that is for the orthogonal system $r_n = [e_n], n \in \mathbb{N}$. (We denote with [ ] the closed linear span).

We only consider the Hilbert cubes whose faces central points are intersections with these rays. Such cubes can be denoted by

$$\{\sigma Q(e_n);\ \sigma \text{ permutation of } \mathbb{N}\}$$

being $\sigma Q(e_n) = \{\bar{x} \in \mathbf{H};\ \bar{x} = \sum x_n e_{\sigma(n)},\ |x_n| \leq 1/n\}$.

Our goal is thus studying operators A of **B** with the condition

$$(*) \qquad \sum \|A\ 1/n\ e_{\sigma(n)}\| < \infty$$

If A verifies (*) we shall say that A gives absolute summability over $r_n \cap \sigma Q(e_n)$. We use the following notation

$$\sum (Q, \sigma, A) = \sum \|A\ 1/n\ e_{\sigma(n)}\|$$

$$\underline{\sum} (Q, A) = \inf \left\{\sum \|A\ 1/n\ e_{\sigma(n)}\|,\ \sigma \text{ permutation of } \mathbb{N}\right\}$$

$$\overline{\sum} (Q, A) = \sup \left\{\sum \|A\ 1/n\ e_{\sigma(n)}\|,\ \sigma \text{ permutation of } \mathbb{N}\right\}$$

1. THE BEHAVIOUR OF OPERATORS WITH RESPECT TO Q

In [Ro.1] the compact operators are characterized by

$$A \in \mathfrak{J} \text{ if and only if } \forall Q,\ \exists \sigma,\ \sum (Q, \alpha, A) < \infty$$

This condition is equivalent to

$$A \in \mathfrak{J} \text{ if and only if } \forall Q,\ \underline{\sum} (Q, A) = 0,$$

which can be proved easily by convenient choices of permutations σ.

In [Ro.2] we have given the following characterization of semi-Fredholm operators

$A \notin \Phi_+$ if and only if $\exists$ Q, $\sum (Q,1,A) < \infty$

(Φ_+ stands for the set of closed range, and finite dimensional kernel operators).

As in the former case it is easily proved

$A \notin \Phi_+$ if and only if $\exists$ Q, $\underline{\sum} (Q,A) = 0$

From now on we shall pay attention to the supremum over the permutations of a given o.n.b.

$$\sup \left\{ \sum \|A\ 1/n\ e_{\sigma(n)}\|,\ \sigma \text{ permutation of } \mathbb{N} \right\}$$

instead of the infimum as we did before.

We give a new characterization of operators which are not semi-Fredholm, by the existence of o.n.b. which make the former supremum arbitrarily small, that is

$$\inf \left\{ \sup \left\{ \sum \|A\ 1/n\ e_{\sigma(n)}\|,\ \sigma \text{ permutation of } \mathbb{N} \right\};\ (e_n) \text{o.n.b. of } \mathbf{H} \right\} = 0,$$

or

$$A \notin \Phi_+ \text{ if and only if } \inf \left\{ \overline{\sum} (Q,A);\ Q \text{ Hilbert cube} \right\} = 0$$

To prove this assertion we have to show first that for a fixed operator A, with non closed range or else with infinite dimensional kernel, and a given $\varepsilon > 0$, there exists a o.n.b. of $\mathbf{H}$, (e_n) verifying

$$\sup \left\{ \sum \|A\ 1/n\ e_{\sigma(n)}\|,\ \sigma \text{ permutation of } \mathbb{N} \right\} < \varepsilon$$

On the other hand if $A \in \Phi_+$, it must be proved that for every o.n.b. (e_n) and for every σ

$$\sum \|A\ 1/n\ e_{\sigma(n)}\| = \infty$$

Along the paper we use several times changes of orthonormal bases in certain finite dimensional subspaces of $\mathbf{H}$. Such changes are expresed in the following form

$$M_p : (e_n;\ n=1,\ldots,2^p) \longrightarrow (u_n;\ n=1,\ldots,2^p)$$

where M_p is a $2^p \times 2^p$ matrix obtained by induction

$$M_1 = \frac{1}{\sqrt{2}} \begin{pmatrix} 1 & 1 \\ 1 & -1 \end{pmatrix} = \frac{1}{\sqrt{2}} N_1$$

$$M_{p+1} = \frac{1}{(\sqrt{2})^{p+1}} \begin{pmatrix} N_p & N_p \\ N_p & -N_p \end{pmatrix}$$

so

$$N_{p+1} = \begin{pmatrix} N_p & N_p \\ N_p & -N_p \end{pmatrix} = \begin{pmatrix} 1 & 1 & & 1 \\ 1 & & & \\ & & \pm 1 & \\ 1 & & & \end{pmatrix}$$

<u>Numerical lemma</u>: a) $\lim_p 2^{-p/2} \sum_{n=1}^{2^p} 1/n = 0$

b) *Let* $(a_n; n \in \mathbb{N})$ *and* $(b_n;\ n \in \mathbb{N})$ *be two non increasing sequences of positive real numbers. Then*

$$\sup \left\{ \sum a_n b_{\sigma(n)};\ \sigma \text{ permutation of } \mathbb{N} \right\} = \sum a_n b_n .$$

<u>Proposition 1.</u> (Operators with non closed range). *Let A be a bounded linear operator with non closed range. Then*

$$\inf\left\{ \bar{\sum}(Q,A);\ Q \text{ Hilbert cube} \right\} = 0$$

Proof. - By [Pl] there exists an o.n.b. of **H**, (e_n) such that

$$\|A e_n\| \longrightarrow 0.$$

Let $\varepsilon>0$. Take a sequence of positive real numbers $(c_n;\ n\in\mathbb{N})$ s.t. $\sum c_n < \varepsilon$. We construct a new o.n.b. of **H** $(v_n; n\in\mathbb{N})$ performing changes in blocks.

Block 1 (B.1). For a given $c_1>0$, $\exists p_1\in\mathbb{N}$ s.t.

$$2^{-p_1/2} \sum_{n=1}^{2^{p_1}} 1/n < \frac{c_1}{2\,\|A\|}$$

which is always possible by the numerical lemma a)

For $p_1\in\mathbb{N}$, $\exists\ a_1\in\mathbb{R}$, $a_1>0$ s.t.

$$a_1 < \frac{\|A\|}{2^{p_1}}$$

Take now $2^{p_1}-1$ elements of the basis (e_n), $\left\{e^1_1, e^1_2, \ldots, e^1_{2^{p_1}-1}\right\}$ such that

$$\|A\, e^1_i\| < a_i,\quad (i=1,\ldots,2^{p_1}-1)$$

(this is always possible since $\|A\, e_n\| \longrightarrow 0$)

Perform the change of o.n.b. in the subspace $[e_1, e^1_1, e^1_2, \ldots, e^1_{2^{p_1}-1}]$, given by

$$M_{p^1}:(e_1,e_1^1,e_2^1,\dots,e^1_{2^{p_1}-1}) \longrightarrow (u_1,\dots,u_{2^{p_1}})$$

Clearly $u_i=2^{-(p_1/2)}(e_1\pm e_1^1\pm e_2^1\pm\dots\pm e^1_{2^{p_1}-1})$ and then $\forall i\in B(1)$,

$$\|A\, u_i\| \le 2^{-(p_1/2)}(\|A\, e_1\|+\|A\, e_1^1\|+\dots+\|A\, e^1_{2^{p_1}-1}\|)$$

$$< 2^{-(p_1/2)}(\|A\, e_1\|+ 2^{p_1} a_1)$$

$$< 2^{-(p_1/2)}(\|A\|+\|A\|) = 2^{-(p_1/2)}\cdot 2\,\|A\|.$$

Call $\{v_1,\dots,v_{2^{p_1}}\}$ the elements $\{u_1,\dots,u_{2^{p_1}}\}$ rearranged so that

$$\|A\, v_1\| \ge \dots \ge \|A\, v_{2^{p_1}}\|$$

Thus

$$\|A\, v_i\| < 2^{-(p_1/2)}\cdot 2\|A\| \quad \forall i\in B(1)$$

and $\displaystyle\sum_{i\in B(1)} (1/i)\ \|A\, v_i\| < 2^{-(p_1/2)}\, 2\|A\| \sum_{i\in B(1)} (1/i)< c_i.$

Let $k_1=\|A\, v_{2^{p_1}}\|$.

Block 2 (B.2). For $c_2>0$, take $p_2\in \mathbb{N}$ s.t.

$$2^{-p_2/2}\sum_{i=1}^{2^{p_2}} 1/i < \frac{\min\{c_2,k_1\}}{2\,\|A\|}$$

and for p_2 choose $a_2 \in \mathbb{R}$, $a_2 > 0$ s.t.

$$a_2 < \frac{\|A\|}{2^{p_2}}$$

In the basis (e_n) drop the elements used in the first block and take $2^{p_2}-1$ elements $\left\{e^2_1, e^2_2, \ldots, e^2_{2^{p_2}-1}\right\}$ s.t.

$$\|A\, e^2_i\| < a_2\,, \quad (i=1,\ldots,2^{p_2}-1)$$

Perform the following change of basis

$$M_{p_2} : (e_2, e^2_1, e^2_2, \ldots, e^2_{2^{p_2}-1}) \longrightarrow (u_{2^{p_1}+1}, \ldots, u_{2^{p_1}+2^{p_2}}).$$

(If e_2 had already been taken in (B.1) we should take the first one not in (B.1)).

$$\text{The } u_i = 2^{-(p_2/2)} (e_2 \pm e^2_1 \pm e^2_2 \pm \ldots \pm e^2_{2^{p_2}-1})$$

$\forall i \in B(2)$, and

$$\|A\, u_i\| \le 2^{-(p_2/2)} (\|Ae_2\| + \|Ae^2_1\| + \|Ae^2_2\| + \ldots + \|Ae^2_{2^{p_2}-1}\| <$$

$$< 2^{-(p_2/2)} . (\|Ae_2\| + 2^{p_2} a_2) < 2^{-(p_2/2)} (\|A\| + \|A\|) =$$

$$= 2^{-(p_2/2)} \cdot 2\, \|A\|.$$

Call now $\left\{v_{2^{p_1}+1}, \ldots, v_{2^{p_1}+2^{p_2}}\right\}$ the rearrangement of $\left\{u_{2^{p_1}+1}, \ldots, u_{2^{p_1}+2^{p_2}}\right\}$ s.t.

$$\|Av_{2^{p_1}+1}\| \ge \ldots \ge \|Av_{2^{p_1}+2^{p_2}}\|.$$

Observe that $\|Av_i\|$ decreases when going from (B.1) to (B.2)

$$\|Av_{2^{p_1}+1}\| < 2^{-(p_2/2)} 2\|A\| < 2^{-(p_2/2)} 2\|A\| \sum (1/i) < k_1 = \|Av_{2^{p_1}}\|.$$

It holds

$$\sum_{i\in B(2)} (1/i)\ \|Av_i\| < 2^{-(p_2/2)}\ 2\|A\| \sum_{i\in B(2)} (1/i) =$$

$$= 2^{-(p_2/2)}\ 2\ \|A\| \left(\frac{1}{2^{p_1}+1} + \ldots + \frac{1}{2^{p_1}+2^{p_2}} \right) <$$

$$< 2^{-(p_2/2)}\ 2\ \|A\|\ (1+1/2+\ldots+1/2^{p_2}) < c_2.$$

Iterating the process we obtain an orthonormal base of H $(v_n; n\in \mathbb{N})$ restricted to the condition

$$\sum (1/n)\ \|Av_n\| < \sum c_n < \varepsilon.$$

Since $(\|Av_n\|;\ n\in\mathbb{N})$ is decreasing, for the Hilbert cube associated to $(v_n;\ n\in\mathbb{N})$,

$$\begin{aligned} \overline{\sum}(Q,A) &= \sup\left\{ \sum \|A\ 1/n\ v_{\sigma(n)}\|\ ,\ \sigma \text{ permutation of } \mathbb{N}\right\} = \\ &= \sup\left\{ \sum 1/n\ \|A\ v_{\sigma(n)}\|\ ,\ \sigma \text{ permutation of } \mathbb{N}\right\} = \\ &= \sum 1/n\ \|A\ v_n\| < \varepsilon\ , \end{aligned}$$

and thus

$$\inf\left\{ \overline{\sum}(Q,A);\ Q \text{ Hilbert cube} \right\} = 0. \quad \square$$

Proposition 2. (Closed range operators with infinite dimensional kernel). *Let A be a bounded linear operator with closed range and infinite dimensional kernel. Then*

$$\inf \left\{ \bar{\sum} (Q, A);\ Q \text{ Hilbert cube} \right\} = 0.$$

Proof.- Distinguish two cases:

Case 1: Codim ker A < ∞.

Equivalently A is a finite rank operator and thus compact. If dim R(A)=k let $(e_n;\ n\in \mathbb{N})$ be an o.n.b. of **H** s.t.

$$\|Ae_k\| \le \ldots \le \|Ae_2\| \le \|Ae_1\| \le \|A\| \quad \text{and} \quad \|Ae_n\|=0 \ ,\forall n>k.$$

Let ε>0, ∃p∈ ℕ s.t.

$$2^{-p/2} \sum_{n=1}^{k2^p} 1/n \ \frac{\varepsilon}{\|A\|} \ .$$

Perform the change of o.n.b.

$$(e_n; n\in \mathbb{N}) \longrightarrow (v_n;\ n\in \mathbb{N}),$$

constructiong $(v_n;\ n\in \mathbb{N})$ by the following blocks:

(B.1) $M_p: (e_1, e_{k+1}, e_{k+2}, \ldots, e_{k+2^p-1}) \longrightarrow (v_1, \ldots, v_{2^p}).$

(B.2) $M_p: (e_2, e_{k+2^p}, \ldots, e_{k+2\cdot 2^p-2}) \longrightarrow (v_{2^p+1}, \ldots, v_{22^p})$

. . . .

(B.k) $M_p: (e_k, \ldots, e_{k\cdot 2^p}) \longrightarrow (v_{(k-1)\cdot 2^p+1}, \ldots, v_{k\cdot 2^p})$ and $v_n = e_n$ $\forall n > K\cdot 2^p$

Then

$$\|Av_i\| = 2^{-(p/2)} \|Ae_i\| \quad \forall i \in (B.1)$$

. . . .

$$\|Av_i\| = 2^{-(p/2)} \|Ae_k\| \quad \forall i \in (B.k)$$

$$\|Av_i\| = 0 \quad \forall i>k2^p$$

Taking into account

$$\|A\| > \|Ae_1\| > \|Ae_2\| > \ldots > \|Ae_k\|,$$

$$\|Av_i\| \leq 2^{-(p/2)} \|A\| \quad \forall i \leq k\cdot 2^p$$

$$\sum_{n\in\mathbb{N}} \frac{1}{n} \|Av_n\| \leq \sum_{n=1}^{k2^p} \frac{1}{n} \|A\| \, 2^{-(p/2)} = 2^{-(p/2)} \|A\| \sum_{n=1}^{k2^p} \frac{1}{n} < \varepsilon.$$

Since $(\|Av_n\|;\ n\in\mathbb{N})$ is non increasing , $\overline{\sum}(Q,A) < \varepsilon$ where Q is the Hilbert cube associated to $(v_n;\ n\in\mathbb{N})$ and $\inf \left\{\overline{\sum}(Q,A); Q \text{ Hilbert cube}\right\} = 0$.

Case 2. Codim ker A=0.

Let $\varepsilon>0$. Take a sequence of positive real numbers $(c_n;\ n\in\mathbb{N})$ s.t. $\sum c_n < \varepsilon$.

Let $(u_n; n \in \mathbb{N})$ and $(e_n;\ n\in \mathbb{N})$ be o.n. bases of ker A and $(\ker A)^{\perp}$ respectively

(B.1). For $c_1>0$ $\exists p_1 \in \mathbb{N}$ s.t.

$$2^{-(p_1/2)} \sum_{n=1}^{2^{p_1}} 1/n < \frac{c_1}{\|A\|}.$$

Let $M_{p_1} : (e_1, u_1, \ldots, u_{2^{p_1}-1}) \longrightarrow (v_1, \ldots, v_{2^{p_1}})$

$$\|Av_i\| = 2^{-(p_1/2)} \|Ae_1\| \ \forall i\in B(1),$$

$$\sum_{i\in(B.1)} \frac{1}{i} \|Av_i\| = \sum_{i\in(B.1)} \frac{1}{i}\, 2^{-(p_1/2)} \|Ae_1\| \leq$$

$$\leq 2^{-(p_1/2)} \|A\| \sum_{i\in(B.1)} \frac{1}{i} < c_1.$$

(B.2) $A\Big|_{(\ker A)^{\perp}}$ is a regular operator and therefore we have the existence of $m, M > 0$ s.t.

$$0 < m < \|Ae_i\| < M < \infty \quad \forall i\in \mathbb{N} .$$

Then $\exists$ a>o s.t.

$$0 < a < \frac{m}{M} < \frac{\|Ae_i\|}{\|Ae_j\|} < \frac{M}{m} < \infty.$$

For $c_2>0$ take $p_2\in \mathbb{N}$, $(p_2>p_1)$ verifying:

a) $2^{-(p_2-p_1)/2} < a$

b) $2^{-(p_2/2)} \displaystyle\sum_{i=2^{p_1}+1}^{2^{p_1+2^{p_2}}} \frac{1}{i} < \frac{c_2}{\|A\|}.$

Observe that b) is possible on account of

$$\sum_{i=2^{p_1}+1}^{2^{p_1}+2^{p_2}} \frac{1}{i} < \sum_{i=1}^{2^{p_2}} \frac{1}{i}$$

By a) we have

$$2^{-(p_2-p_1)/2} < a = \frac{m}{M} < \frac{\|Ae_i\|}{\|Ae_j\|} \quad \forall i, j \in \mathbb{N}$$

and so

$$2^{-(p_2/2)} \|Ae_2\| < 2^{-(p_1/2)} \|Ae_1\|$$

which ensures that $\|Av_i\|$ is decreasing when going from the first block to the second one.

Iterating the process

$$\sum 1/n \|A v_n\| < \sum c_n < \varepsilon$$

and

$$\inf \left\{ \bar{\sum} (Q, A);\ Q \text{ Hilbert cube} \right\} = 0. \qquad \square$$

Proposition 3. (Closed range operators with finite dimensional kernel. Semi-Fredholm operators Φ_+). *Let* A *be a bounded linear operator with closed range and finite dimensional kernel. Then*

$$\forall\ (e_n;\ n\in \mathbb{N}) \text{ o.n.b } of\ \ H\ ,\ \sum 1/n\ \|Ae_n\| = \infty$$

and so

$$\inf \left\{ \bar{\sum} (Q, A);\ Q \text{ Hilbert cube} \right\} = \infty,$$

Proof. It is clear when A is injective, since then for every o.n.b. $(e_n;\ n\in \mathbb{N})$, $(\|Av_n\|;\ n\in \mathbb{N})$ is bounded from above and from below.

For $0\neq\dim \operatorname{Ker} A < \infty$, let $S=(e_n;n\in\mathbb{N})$ be an o.n.b. and S' the sequence obtained from S dropping -if there are any- the elements e_i which are also in Ker A. (They are at most finitely many).

Then $S'\cap \ker A= \emptyset$, and by [P1]

a) $\inf \left\{\|Ae_i\|;\ e_i\in S'\right\} = 0$

is equivalent to

b) $\exists(u_n;\ n\in\mathbb{N})$ o.n.b. of $(\ker A)^{\perp}$ s.t.

$$\inf \left\{ \|Au_n\|;\ n\in \mathbb{N} \right\} = 0.$$

But if b) holds, $A|_{(\ker A)^{\perp}}$ must be a non closed range operator and so would be A, which leads to a contradiction.

So $\inf \left\{ \|Ae_i\|;\ e_i\in S' \right\} > 0$ and $\sum 1/n\ \|Ae_n\| = \infty$. □

REFERENCES.

[Bu] Burillo, P.J. (1976). Una conjetura de Maurin sobre operadores nucleares en el espacio de Hilbert. Collect. Math., T. 27.

[Go-K] Gohberg, I.C.; Krejn, M.G. (1971). Introduction a la théorie des opérateurs non auto-adjoints dans un espace hilbertien. Dunod.

[Go-Ma] Gohberg, I.C. ; Markus, A.S. (1964). Some relations between eigenvalues and matrix elements of linear operators. Mat. Sbornik 64 (106), p. 481-496.

[Ho] Holub, J.R. (1971). Characterization of nuclear operators in Hilbert space. Rev. Roumaine Math. Pures et Appl. T. XVI n. 5.

[Pl] Plans, A. (1980). Comportamiento de los operadores lineales acotados en sistemas ortogonales. Acatas de las VII Jornadas Mat. Hispano-Lusas (1980).

[Ro1] Rodes, A.A. (1980). Ideales de operadores lineales acotados y sumabilidad en el espacio de Hilbert. Tesis Doctoral. Zaragoza.

[Ro2] Rodes, A.A. (1984). Una caracterización de los operadores de Fredholm y Semi-Fredholm. Rev. de la Academia de Ciencias de Madrid. T. LXXVIII p. 497-501.

OPERATORS FROM H^1 INTO A BANACH SPACE AND VECTOR VALUED MEASURES.

Oscar Blasco
Departamento de Matemáticas, Univ. de Zaragoza, Spain.

A mi Profesor Antonio Plans.

Abstract. The space of linear operators from the Hardy space H^1, defined by atoms, into a Banach space is characterized in terms of a class of vector valued measures of bounded mean oscillation.

1. INTRODUCTION AND PREVIOUS DEFINITIONS.

The objective of this paper is to give a representation of operators from the "atomic" Hardy space H^1 into a Banach space using vector valued measures. In [7] the reader can see a representation of operators in $\mathcal{L}(L^p,X)$ in terms of X-valued measures of bounded p'-semivariation. Here we shall introduce a class of vector valued measures closely related to functions of bounded mean oscillation. We shall obtain also some results of duality and others concerning the Radon-Nikodym property.

Throughout the paper $(X,\|\cdot\|_X)$ will be a Banach space and $(\mathbb{T},\mathcal{B},m)$ will denote the Lebesgue space on the circle $\mathbb{T}$ with normalized measure. The reader is referred to ([2], [5]) for notions and results concerning Hardy spaces, and to ([3],[4]) for those related to vector measures and vector valued functions.

Let us recall some definitions we shall use later on.

Definition 1. (See [2],[5]) A fuction a in $L^1(\mathbb{T})$ is said to be an atom if

i) supp $a \subset I$, where I is an interval.

ii) $\int_I a(t)\, dm(t) = 0$

iii) $|a(t)| \le m(I)^{-1}$ for all $t \in I$

The constants $a(t)=z$ with $|z|=1$ are also considered as atoms. The space H^1 is now defined as follows

$$(1.1) \qquad H^1 = \left\{ f \in L^1(\mathbb{T}) : f = \sum \lambda_k a_k,\ a_k \text{ are atoms and } \sum |\lambda_k| < \infty \right\}$$

The norm in this space is given by

$$(1.2) \qquad \|f\|_{at} = \inf \left\{ \sum |\lambda_k| : f = \sum \lambda_k a_k ,\ a_k \text{ atoms} \right\}$$

Two very well known facts we shall use later are the density of the simple functions in H^1 and the fact that the series $f = \sum \lambda_k a_k$ converges in $L^1(\mathbb{T})$.

Now inspired by BMO space of John-Nirenberg [6] we define a new class of vector valued meausres.

Given a X-valued measure G, let us denote by $\|G\|(E)$ the semivariation of G over the set. E, i.e.

$$\|G\|(E) = \sup \left\{ |\xi\, G|(E) : \|\xi\|_{X^*} = 1 \right\}$$

where ξG represents the complex measure $\xi G(A) = \langle \xi, G(A) \rangle$ and $|\xi G|(E)$ stands for the variation over E.

<u>Definition</u> 2. Let I be an interval and G a X-valued measure of bounded semivariation. We define the following measure

$$(1.3) \qquad G^*_I(E) = G(E \cap I) - (G(I)/m(I))m(E \cap I)$$

We shall say that G belongs to $w\mathcal{BMO}(X)$ if

$$(1.4) \qquad |G|^* = \sup \left\{ m(I)^{-1} \|G^*_I\|(I) : I \text{ interval} \right\} < +\infty.$$

In [1] the author considered a stronger class of vector valued measures, denoted by $\mathcal{BMO}(X)$, where the semivariation is replaced by the variation of the measures. This space was used to characterize the dual space of the atomic Hardy space of vector valued functions.

Observe that $G(E) = xm(E)$ for some vector x satisfies tha $|G|^* = 0$. Therefore, in order to get a norm, we consider

$$(1.5) \qquad |G|_{w\mathcal{BMO}(X)} = |G|^* + \|G(\Pi)\|_X$$

There are special measures in $w\mathcal{BMO}(X)$ coming from X-valued functions.

Definition 3. Let f be a X-valued Pettis integrable function. We shall say that f belongs to wBMO(X) if

$$(1.6) \qquad |f|^* = \sup \left\{ m(I)^{-1} \int_I |\xi f(t) - (\xi f)_I| dm(t) \right\} < +\infty$$

where the sup is taken over all interval I and ξ in X^*, and $(\xi f)_I$ stands for $m(I)^{-1} \int_I \langle \xi, f(t) \rangle \, dm(t)$.

For similar reasons as above we define

$$(1.7) \qquad |f|_{wBMO(X)} = |f|^* + \|(P) \int f(t) \, dm(t)\|_X .$$

To connect functions and measures let us recall that if f is a X-valued Pettis integrable function with $\sup \left\{ \|\xi f\|_1 : \|\xi\|_{X^*} = 1 \right\} < +\infty$ then

$$(1.8) \qquad F(E) = (P) \int_E f(t) \, dm(t)$$

defines a X-valued measure of bounded semivariation, and under this identification we can write

(1.9) $\quad wBMO(X) \subset w\mathcal{BMO}(X)$ (isometric inclussion).

The coincidence between both spaces will be guaranteed only for spaces with the Radon-Nikodym property as we shall see in Corollary 3.

2. THE THEOREM.

The lemma we shall need uses an idea coming from martingale theory and allows us to associate to each vector valued measure in $w\mathcal{BMO}(X)$ a sequence of vector valued functions contained within a ball of $wBMO(X)$.

Denoting by $I_{n,k} = (k-1)2^{-n}, k2^{-n})$ the dyadic intervals of length 2^{-n}, and taking a X-valued measure G we define

$$g_n = \sum_{k=0}^{2^n-1} x_{n,k}\, \chi_{I_{n,k}} \tag{2.1}$$

where $x_{n,k} = G(I_{n,k})/m(I_{n,k})$.

The proof of the following lemma can be done by reproducing the one done in [1] for the space $\mathcal{BMO}(X)$ instead of $w\mathcal{BMO}(X)$.

<u>Lemma</u>.- *Let* G *belong to* $w\mathcal{BMO}(X)$ *then*

$$\|g_n\|_{wBMO(X)} \le 4\, \|G\|_{w\mathcal{BMO}(X)} \tag{2.2}$$

(2.3) $\quad \int_E g_n(t)\, dm(t)$ *converges to* $G(E)$ *as* $n\to\infty$ *for all* E *in* $\mathcal{B}$.

<u>Theorem</u>.- $w\mathcal{BMO}(X) = \mathcal{L}(H^1, X)$ *(with equivalent norms).*

<u>Proof</u>. Let us consider T belonging to $\mathcal{L}(H^1, X)$ and define the vector measure

$$G(E) = T(\aleph_E) \tag{2.4}$$

To show that G belongs to $wBMO(X)$, let us take an interval I and observe that

$$G^*_I(E) = T(\aleph_{E\cap I} - m(E\cap I)m(I)^{-1}\aleph_I) = 2m(I)T(a_E)$$

where $a_E = (2m(I))^{-1}(\aleph_{E\cap I} - m(E\cap I)m(I)^{-1}\aleph_I)$ is an atom. Therefore $\|G^*_I(E)\|_X \le 2m(I)\|T\|$. Now, using the fact that

$$\|G^*_I\|(I) \le 4 \sup\left\{ \|G^*_I(E)\| : E\subset I\right\} \text{ (see [3], page 4)}$$

then we have $\|G\|_{wBMO(X)} \le 8\|T\|$.

Conversely let us take G in $wBMO(X)$ and define the operator from the simple functions into X given by

$$(2.5) \qquad T_G\left[\sum_{j=1}^m \lambda_j \chi_{E_j}\right] = \sum_{j=1}^m \lambda_j\, G(E_j)$$

By the lemma we can consider the sequence g_n in $L^\infty(X)$ and the corresponding operators T_n in $\mathcal{L}(L^1(\mathbb{T}),X)$

$$(2.6) \qquad T_n(\varphi) = \int \varphi(t)\, g_n(t)\, dm(t)$$

Notice that for an atom a supported in I we can write

$$\begin{aligned}\|T_n(a)\|_X &= \left\|\int a(t)g_n(t)dm(t)\right\|_X \\ &= \sup\left\{\left|\int_I a(t)\ \langle\xi, g_n(t)\rangle dm(t)\right| : \|\xi\|_{X^*} = 1\right\} \\ &= \sup\left\{\left|\int_I a(t)\ (\xi g_n(t) - (\xi g_n)_I) dm(t)\right| : \|\xi\|_{X^*} = 1\right\} \\ &\le \sup\left\{ m(I)^{-1} \int_I |\xi g_n(t) - (\xi g_n)_I|dm(t) : \|\xi\|_{X^*} = 1\right\} \\ &\le |g_n|^* \le 4\|G\|_{wBMO(X)}.\end{aligned}$$

Since for constant atoms we can show $\|T_n(a)\|_X \le \|G\|_{wBMO(X)}$, then

(2.7) $$\|T_n(a)\|_X \le 4\ \|G\|_{w\mathcal{BMO}(X)} \quad \text{for all atom}$$

Now using that T_n belongs to $\mathcal{L}(L^1(\mathbb{T}),X)$ and that for each function f in H^1 any representation $f=\sum \lambda_k a_k$ converges in $L^1(\mathbb{T})$, (2.7) implies

(2.8) $$\|T_n(\varphi)\|_X \le 4\ \|G\|_{w\mathcal{BMO}(X)} \cdot \|\varphi\|_{at} \quad \text{for all } \varphi \text{ in } H^1.$$

Now combining (2.8) and (2.3) we can easily prove that

$$\|T(s)\|_X \le 4\ \|G\|_{w\mathcal{BMO}(X)} \cdot \|s\|_{at} \quad \text{for all simple functions}$$

what allows us to extend the operator to an element in $\mathcal{L}(H^1,X)$ finishing the proof.

It is not hard to show that for $X = \mathbb{C}$ all spaces considered here and in [1] coincide, that is

$$wBMO(\mathbb{C}) = w\mathcal{BMO}(\mathbb{C}) = \mathcal{BMO}(\mathbb{C}) = BMO$$

<u>Corollary 1</u>.- $(H^1)^* = BMO$.

<u>Corollary 2</u>.- $(H^1 \hat{\otimes} X)^* = w\mathcal{BMO}(X^*)$

<u>Proof</u>. It follows from the identification $(A \hat{\otimes} B)^* = \mathcal{L}(A,B^*)$ (see [3], page 230).

<u>Corollary 3</u>.- *If* $wBMO(X) = w\mathcal{BMO}(X)$ *then* X *has the Radon-Nikodym property.*
<u>Proof</u>. Use that $\mathcal{L}(L_1(\mathbb{T}),X) \subset \mathcal{L}(H^1,X)$ and the characterization of RNP in terms of representability of operators (see [3], page 63).

REFERENCES.

[1] Blasco, O. Hardy spaces of vector valued functions: duality. To appear in Trans. Amer. Math. Soc.
[2] Coifman, R.R. & Weiss, G. (1977). Extension of Hardy spaces and their use in Analysis, Bull. Amer. Math. Soc. 83, 560-645.
[3] Diestel, J. & Uhl, J.J. (1977).Vector Measures. Math. Surveys, 15, American Mathematical Society, Providence, Rhode Island.
[4] Dinculeanu, N. Vector measures (1967). Pergamon Press, New York.
[5] Garcia-Cuerva, J. & Rubio de Francia, J.L.(1985). Weigthed norm inequalities and related topics, North-Holland, Amsterdam
[6] John, F. & Nirenberg. L. (1961). On functions of bounded mean oscillation, Comm. Pure Appl. Math. 14, 415-426.
[7] Phillips, R. (1940). On linear transformations, Trans. Amer. Math. Soc. 48, 516-541.

Supported by the Grant C.A.I.C.Y.T. PB 85-0338.

AMS Classification (1980): 46G10, 42B30, 46B22.

Key words: Vector-valued measures, bounded mean oscillation, atom.

OPERATORS ON VECTOR SEQUENCE SPACES.

Fernando Bombal
Universidad Complutense de Madrid, Spain.

Dedicated to Antonio Plans

Abstract.- A representation theorem for bounded linear operators on spaces $(\Sigma\oplus E_n)_p$ in terms of induced operators on each E_n is established, and some particular classes of operators are characterized in this way. Applications to the study of the structure of $(\Sigma\oplus E_n)$ are given.

INTRODUCTION AND NOTATIONS.

Let $E_n (n\in \mathbb{N})$ be a sequence of Banach spaces over $\mathbb{K}$ ($\mathbb{R}$ or $\mathbb{C}$), and $1\le p<\infty$. We shall denote, as usual, by $(\Sigma\oplus E_n)_p$ the space of all vector valued sequences $x=(x_n)$ such that $x_n\in E_n$ $(n=1,2,\dots)$ and $\|x\|_p^p = \sum_{n=1}^{\infty} \|x_n\|^p$ is finite, endowed with the Banach norm $x\longrightarrow\|x\|_p$. Similarly, we define the c_0-sum $(\Sigma\oplus E_n)_0$ and ℓ_∞-sum $(\Sigma\oplus E_n)_\infty$. It is well known that if $E = (\Sigma\oplus E_n)_p$ $(1\le p<\infty$ or $p=0)$, then E^*, the topological dual of E, can be identified to $E=(\Sigma\oplus E_n)_q$, where q is the number conjugat9ed to p, that is, $\frac{1}{p} + \frac{1}{q} = 1$ if $1\le p<\infty$ or $p=0$, where $\frac{1}{0} = \infty$ and $\frac{1}{\infty} = 0$ as usual. This identification is given by de isometry

$$(\Sigma\oplus E_n^*)_q \ni x^*=(x_n^*)\longrightarrow T_{x^*} \in (\Sigma\oplus E_n)_p^* , \quad T_{x^*}(x)= \sum_{n=1}^{\infty} \langle x_n, x_n^*\rangle.$$

For every $m\in \mathbb{N}$, I_m will denote the canonical injection $E_m\ni y \longrightarrow (0,\dots,0,\overset{(m}{y},0,\dots)\in (\Sigma\oplus E_n)_p$ and Π_m the canonical projection

$(\Sigma\oplus E_n)_p \ni x=(x_n)\longrightarrow x_m \in E_m$. If F is another Banach space over K and $T:(\Sigma\oplus E_n)_p \longrightarrow F$ is an operator (i.e., a continuous linear map), then for every $n\in \mathbb{N}$, $T_n=T\cdot I_n$ is an operator from E_n into F. In this note, we characterize the operators on $(\Sigma\oplus E_n)_p$ in terms of the T_n's, giving some applications to the study of the structure of this space.

Given the Banach spaces E and F, we shall write L(E,F) to denote the space of all operators between E and F, and B(E) will stand for the closed unit ball of E. The rest of notations and terminology used and not defined along the paper, will be the usual in Banach space theory, as can be seen for example in [5] and [8].

1. OPERATORS ON $(\Sigma\oplus E_n)_p$.

Theorem 1.1. *Let* E_n $(n\in \mathbb{N})$, F *be Banach spaces and* $1\le p<\infty$ *or* $p=0$. *Every operator* T *from* $(\Sigma\oplus E_n)_p$ *into* F *determines a unique sequence* (T_n) *of operators, where* $T_n\in L(E_n,F)$ *such that*

I. For every $y^*\in F^*$, $(y^*T_n)=(T_n^*(y^*))\in (\Sigma\oplus E_n^*)_q$, *being* q *conjugated to* p.

II. The set $\left\{(T_n^*(y^*))_{n=1}^{\infty} : y^*\in B(F^*)\right\}$ *is bounded in* $(\Sigma\oplus E_n^*)_q$

Moreover,

III. $T(x)=\sum_{n=1}^{\infty} T_n(x_n)$ *for every* $x=(x_n)\in (\Sigma\oplus E_n)_p$.

and

IV. $\|T\| = \sup\left\{\|(T_n^*(y^*))\|_q : y^*\in B(F^*))\right\}$.

Conversely, if (T_n) *is a sequence of operators with each* T_n *belonging to* $L(E_n,F)$, *such that (I) and (II) are satisfied, then (III) defines an operator* T *fron* $(\Sigma\oplus E_n)_p$ *to* F *verifying (IV).*

Proof. Let T be an operator from $(\Sigma\oplus E_n)_p$ into F and define $T_n = T\cdot I_n \in$ $\in L(E_n,F)$ Every x in $(\Sigma\oplus E_n)_p$ can be written as a convergent series

$$x = \sum_{n=1}^{\infty} I_n\Pi_n(x),$$

and so

$$T(x) = \sum_{n=1}^{\infty} T_n \Pi_n(x) \text{ in } F,$$

which is (III). On the other hand, if $y^* \in F^*$ then

$$T^*(y^*) \in (\Sigma \oplus E_n)_p^* = (\Sigma \oplus E_n^*)_q ,$$

with the canonical identification (see the introduction). If $T^*(y^*)=(x_n^*)$, where $x_n^* \in E_n^*$, then for every $x=(x_n) \in (\Sigma \oplus E_n)_p$ we have

$$\sum_{n=1}^{\infty} \langle x_n, x_n^* \rangle = \langle T(x), y^* \rangle = \sum_{n=1}^{\infty} \langle T_n(x_n), y^* \rangle ,$$

which proves that $x_n^* = T_n^*(y^*)$ for every $n \in \mathbb{N}$, and so we get (I). The continuity of T^* proves then (II). Finally,

$$\|T\| = \|T^*\| = \sup \left\{ \|T^*(y^*)\| : y^* \in B(F^*) \right\} = \sup \left\{ \|(T_n^*(y^*))\|_q : y^* \in B(F^*) \right\},$$

which is (IV).

For the converse, let us consider only the case $1<p<\infty$, being obvious the modifications needed when p is 0 or 1. Let us then suppose that for every n we have $T_n \in L(E_n, F)$ satisfying (I) and (II). Let $x=(x_n) \in (\Sigma \oplus E_n)_p$, $\varepsilon>0$, $M = \sup \left\{ \|(T^*(y^*))\| : y^* \in B(F^*) \right\}$ and $n_0 \in \mathbb{N}$ such that $\left[\sum_{n \geq n_0} \|x_n\|^p \right]^{1/p} < \varepsilon/M$. Then, if $r > s \geq n_o$ we have

$$\| \sum_{n=s}^{r} T_n(x_n) \| = \operatorname{Sup} \left\{ | \sum_{n=s}^{r} \langle T_n(x_n), y^* \rangle | : y^* \in B(F^*) \right\} \leq$$

$$\leq \operatorname{Sup} \left\{ \left[\left(\sum_{n=s}^{r} \|T_n^*(y^*)\|^q \right)^{1/q} \right] \left[\left(\sum_{n=s}^{r} \|x_n\|^p \right)^{1/q} \right] : y^* \in B(F^*) \right\} \leq \varepsilon.$$

Thus, $\Sigma T_n(x_n)$ is Cauchy in F, and so converges. Let us define T by (III).

Clearly T is linear and

$$|\langle T(x), y^*\rangle| \le M\|x\|_p \ , \ \forall y^* \in B(F^*),$$

which proves that T is continuous and $\|T\| \le M$. From the definition we get $T^*(y^*) = (T_n^*(y^*))$ and reasoning as in the first part of the proof we obtain (IV).

Let us consider now some special classes of operators. For a pair of Banach spaces E and F we shall write:

- $K(E,F)=\{T\in L(E,F)$: T(B(E)) is relatively compact$\}$ (compact operators).
- $\omega(E,F)=\{T\in L(E,F)$: T(B(E)) is weakly relatively compact$\}$ (weakly compact operators).
- $D(E,F)=\{T\in L(E,F)$: T sends weakly Cauchy sequences into weakly convergent sequences$\}$ (Dieudonné operators).
- $DP(E,F)=\{T\in L(E,F)$: T sends weakly Cauchy sequences into norm convergent ones$\}$ (Dunford-Pettis operators).
- $U(E,F) = \{ T \in L(E,F)$: T sends weakly Cauchy sequences into unconditionally convergent ones $\}$ (unconditionally converging operators).
- $S(E,F)=\{T\in L(E,F)$: For any infinite dimensional subspace M of E, T restricted to M is not a topological isomorphism$\}$ (Strictly singular operators).

Each of the above mentioned classes of operators are closed operator ideals in the sense of Pietsch, that is, they form a closed vector subspace of $L(E,F)$ and are stable under both side composition with bounded linear maps (See [8]). Also notice that the following inclusions hold

$$(*) \qquad K(E,F) \begin{matrix} \subset \ \omega(E,F) \ \supset \\ \supset \ DP(E,F) \ \subset \end{matrix} D(E,F) \subset U(E,F) \subset L(E,F)$$

$$(**) \qquad K(E,F) \subset S(E,F) \subset U(E,F).$$

According to theorem 1.1., if $T:(\Sigma\oplus E_n)_p \longrightarrow F$ belongs to some operator ideal ϑ, clearly all the operators in the representing sequence

(T_n) of T belong to the same operator ideal. The converse is not true in general, as the following examples show.

Example 1.2. For each $n \in \mathbb{N}$, let us take a Banach space $E_n \neq \{0\}$ of finite dimension, and let I be the identity operator on $(\Sigma \oplus E_n)_p$. Then, each I_n is a finite range operator, and so it belongs to all the classes of operators considered, but:

a) I is an isomorphism. Hence, it is neither compact nor strictly singular.

b) When p=1, $(\Sigma \oplus E_n)_1$ contains a copy of ℓ_1, and so I is not weakly compact.

c) When p=0, $(\Sigma \oplus E_n)_0$ contains a copy of c_0, and so I is not unconditionally converging.

d) When p>1, $(\Sigma \oplus E_n)_p$ is reflexive. Hence, I is not Dunford-Pettis, because any Dunford-Pettis operator on a space that contains no copy of ℓ_1, is compact by Rosenthal's ℓ_1-theorem, (see, f.i [5] 2.e.5).

In order to obtain some results in the positive direction, we need the following lemmas:

Lemma 1.3. *Let* E_n $(n \in \mathbb{N})$, F *be Banach spaces,* $1 \leq p < \infty$ *and* T *an operator from* $E = (\Sigma \oplus E_n)_p$ *into* F *with representing sequence* (T_n). *Suppose* $A \subset E$ *satisifes the following condition*

(+) *For every* $\varepsilon > 0$, *there exist* $n_\varepsilon \in \mathbb{N}$ *such that*

$$\sup_{x \in A} \sum_{n \geq n_\varepsilon} \|\Pi_n(x)\|^p < \varepsilon$$

Then

$$\lim_{m \to \infty} \sup_{x \in A} \left\| \sum_{n=1}^{m} T_n \Pi_n(x) - T(x) \right\| = 0$$

Proof. Let $\varepsilon > 0$ and choose m_0 such that

$$\sup_{x \in A} \sum_{n \geq m_0} \|\Pi_n(x)\|^p \leq (\varepsilon / \|T\|)^p$$

Then, if $y^* \in B(F^*)$ and $m>m_0$ for every $x \in A$ we have

$$|< \sum_{n=1}^{m} T_n \Pi_n(x) - T(x), y^* >| = | \sum_{n>m} <\Pi_n(x), T_n^*(y^*)>| \le$$

$$\le \left[\sum_{n>m} \|\Pi_n(x)\|^p\right]^{1/p} \left[\sum_{n>m} \|T_n^*(y^*)\|^q\right]^{1/q} \le (\varepsilon/\|T\|)\cdot\|T\| = \varepsilon,$$

by (IV) of th. 1.1 the result follows, taking the supremum when $y^* \in B(F^*)$.

We also need the following well known result:

Lemma 1.4. ([2], Lemma 6). *Let* H *be a subset of a Banach space. If for every* $\varepsilon>0$ *there exists a (weakly) compact subset* H_ε *of* E *such that* $H \subset H_\varepsilon + \varepsilon B(E)$, *then* H *is relatively (weakly) compact.*

Theorem 1.5. *Let* E_n $(n \in \mathbb{N})$, F *be Banach spaces,* $1 \le p < \infty$ *and* T *an operator from* $E = (\Sigma \oplus E_n)_p$ *into* F *with representing sequence* (T_n). *Then*

a) If $p=1$, T *is unconditionally converging (resp. Dieudonné, Dunford-Pettis), if and only if so is each* T_n.

b) If $p>1$, T *is weakly compact, if and only if so is each* T_n.

Proof.

a) Let us consider first the case of unconditionally converging operators. Suppose each T_n is unconditionally converging and let Σx^k $(x^k = (x_n^k) \in E)$ be a weakly unconditionally Cauchy series in E. It suffices to show that $\|T(x^k)\|$ tends to zero. As $(T(x^k))$ is weakly null, we only need to show that $H = \{T(x^k): k \in \mathbb{N}\}$ is norm relatively compact. In the first place, let us notice that the set $H_n = \{T_n(\Pi_n(x^k)): k \in \mathbb{N}\}$ is relatively norm compact for each $n \in \mathbb{N}$ by hypothesis. Reasoning as in the case $E_n = \mathbb{K}$ for every n, one can easily prove that the set $A = \{x^k: k \in \mathbb{N}\}$, being relatively weakly compact, satisfies condition (+) of lemma 1.3. Then, given $\varepsilon>0$, there exists $m \in \mathbb{N}$ such that $\|\sum_{n>m} T_n \Pi_n(x^k)\| < \varepsilon$ for every k, i.e.

$$T(x_k) \in H_\varepsilon + \varepsilon B(F), \text{ for every } k \in \mathbb{N},$$

where H_ε is the norm closure of $H_1+\ldots+H_m$, a norm compact set. Lemma 1.4 applies, giving the result.

The proof for Dieudonné and Dunford-Pettis operators is similar: let (x^k) be weakly Cauchy in E. Since $(T(x^k))$ is weakly Cauchy, to prove that it is weakly (resp. norm) convergent, we only need to show that $H = \{T(x^k): k\in\mathbb{N}\}$ is relatively weak (resp. norm) compact, and this can be proved as before.

Let us remark that another proof for the case of Dunford-Pettis operators is included in [3], Prop. 14.3.

b) See [3], prop. 14.4.

Corollary 1.6. *Let ϑ be an operator ideal. Then, with the notations of theorem 1.5, we have*

a) $\vartheta((\Sigma\oplus E_n)_1,F) \subset \psi((\Sigma\oplus E_n)_1,F)$ *if and only if* $\vartheta(E_n,F) \subset \psi(E_n,F)$ *for every* $n\in\mathbb{N}$ *(where ψ = DP, U or D).*

b) For $1<p<\infty$, $\vartheta((\Sigma\oplus E_n)_p,F) \subset \omega((\Sigma\oplus E_n)_p,E)$ *if and only if* $\vartheta(E_n,F) \subset \omega(E_n,F)$ *for every* $n\in\mathbb{N}$.

Proof. In one sense, the proof follows directly from the preceding theorem. For the converse it suffices to take into account that for every $S\in\vartheta(E_n,F)$, $S\cdot\Pi_n\cdot I_n = S$ *and* $S\cdot\Pi_n\in\vartheta((\Sigma\oplus E_n)_p,F)$.

When p=0, we can sharpen considerably the above results, as the following theorem shows.

Theorem 1.7. *Let ϑ be a closed operator ideal contained in U. Let E_n ($n\in\mathbb{N}$), F be Banach spaces and T an operator from $E=(\Sigma\oplus E_n)_0$ into F, with representing sequences (T_n). Then, the following assertions are equivalent:*

a) $T \in \vartheta(E,F)$

b) $T_n\in\vartheta(E_n,F)$ *for every* $n\in\mathbb{N}$, *and*

$$\lim_{m\to\infty} \left\| \sum_{n=1}^{m} T_n\cdot\Pi_n - T\right\| = 0.$$

Proof.

(a)⇒(b) Obviously, $T_n\in\vartheta(E_n,F)$ for every $n\in\mathbb{N}$. Let us prove the second assertion. If $\|\sum_{n=1}^{m} T_n\cdot\Pi_n - T\|$ does not converge to 0, then there exists an $\varepsilon>0$ and, for every m, a q>m such that $\|\sum_{n=1}^{m} T_n\cdot\Pi_n - T\| > \varepsilon$. Take q_1 and $x^1\in E$ such that $\|x^1\|\le 1$ and

$$\|\sum_{n=1}^{q_1} T_n\cdot\Pi_n(x^1)-T(x^1)\| = \|\sum_{n>q_1} T_n\cdot\Pi_n(x^1)\| > \varepsilon.$$

There is a $p_1>q_1$ such that

$$\|\sum_{n=q_1+1}^{p_1} T_n\cdot\Pi_n(x^1)\| > \varepsilon.$$

Take $q_2>p_1$ such that $\|\sum_{n=1}^{q_2} T_n\cdot\Pi_n - T\| > \varepsilon$ and proceed inductively. Then we get a sequence (x^n) in B(E) and a sequence of natural numbers $q_1<p_1<q_2<\dots$ such that

$$\|\sum_{n=q_j+1}^{p_j} T_n\cdot\Pi_n(x^j)\| > \varepsilon, \quad j=1,2,\dots$$

Put $y^j = \sum_{n=q_j+1}^{p_j} T_n\cdot\Pi_n(x^j)$; then $\|y^j\|\le 1$ for every j and, as the y^j's have disjoint supports, the series $\sum y^j$ is weakly unconditionally Cauchy in E. However,

$$\|T(y^j)\| = \|\sum_{n=q_j+1}^{p_j} T_n\cdot\Pi_n(x^j)\| > \varepsilon.$$

which contradicts the fact that T is unconditionally converging.

(b)⇒(a) As $\sum_{n=1}^{m} T_n \cdot \Pi_n \in \vartheta(E,F)$ for every m, it follows inmediately from the fact that ϑ is closed.

Corollary 1.8. *Let ϑ_1, ϑ_2 be closed operator ideals, both contained in U, the unconditionally converging operators. Then with the notations of theorem 1.7, the following assertions are equivalent:*

a) $\vartheta_1(E,F) \subset \vartheta_2(E,F)$

b) $\vartheta_1(E_n,F) \subset \vartheta_2(E_n,F)$ *for every* $n \in \mathbb{N}$.

Proof. It is completely analogous to that of corollary 1.6, using theorem 1.7 instead of theorem 1.6.

2. SOME APPLICATIONS.

Throughout this section, E_n ($n \in \mathbb{N}$) will be Banach spaces. Many important properties of a Banach space are (or can be) defined in terms of the behaviour of some classes of operators on it. Let us recall some of them: A Banach space is said to have

- The Dunford-Pettis property (DPP in short) if

  $\omega(E,.) \subset DP(E,.)$

- The Dieudonné property (DP in short) if $\omega(E,.) = D(E,.)$.
- The reciprocal Dunford-Pettis property (RDPP in short) if

  $DP(E,.) \subset \omega(E,.)$.

- The Grothendieck's property (GP in short) if

  $L(E,c_0) = \omega(E,c_0)$

- The Pelczynski's V property (VP in short) if

  $\omega(E,.) = U(E,.)$.

The four first properties were introduced by Grothendieck in [4] and the last one by Pelczynski in [6]. These properties are invariant by isomorphism and are inherited by finite products and complemented subspaces. In consequence, if $(\Sigma\oplus E_n)_p$ has some of the above mentioned properties, the same happens to each E_n, which is isomorphic to a complemented subspace of $(\Sigma\oplus E_n)_p$. The results obtained in section 1 allow us to get some information in the opposite direction. The following first three propositions contain eventually the results of Chapters 13 and 14 in [3], where a different proof is given:

Proposition 2.1. ([3], ch. 13) $(\Sigma\oplus E_n)_0$ *has the* DPP *(resp.* RDPP, DP, VP) *if and only if so does every* E_n.

Proof. It follows from corollary 1.8.

Proposition 2.2. ([3], 14.3) $(\Sigma\oplus E_n)_1$ *has the* DPP *if and only if so does each* E_n.

Proof. Follows inmediately from corollary 1.6(a).

Proposition 2.3. ([3], 14.4) *Let* $1<p<\infty$. *Then* $(\Sigma\oplus E_n)_p$ *has the* RDPP *(resp.* DP, VP*) if and only if so does every* E_n.

Proof. Follows from corollary 1.6(b).

Remark 2.4.

a) $(\Sigma\oplus E_n)_1$ has a complemented subspaces isomorphic to ℓ_1, and so it can not have either the DP, the RDPP or the VP. In the same way $(\Sigma\oplus E_n)_p$ $(p>1)$ contains a complemented copy of ℓ_p and so it can not have the DPP.

b) In [1] are defined and studied the properties corresponding to the relations $D(E,.) = DP(E,.)$ and $U(E,.) = DP(E,.)$. Obviously, corollary 1.6(a) yields a result similar to proposition 2.2. for these properties.

Proposition 2.5. *let* $1<p<\infty$. *Then*

a) $(\Sigma\oplus E_n)_p$ *has the* GP *if annd only if every* E_n *has it.*

b) $(\Sigma\oplus E_n)_p$ *has a complemented copy of* ℓ_1 *if and only if there exists* $n\in\mathbb{N}$ *such that* E_n *contains a complemented copy of* ℓ_1.

Proof. (a) and (b) follow directly from corollary 1.6(b), taking into account the following result of Rosenthal [9]: Let T be an operator from a Banach space E into ℓ_1. Then either T is weakly compact or E contains a complemented copy X of ℓ_1 such that the restriction $T|_X$ is an isomorphism. In consequence, we have the followig operator characterization:

E contains a complemented copy of ℓ_1 if and only if $L(E,\ell_1) = \omega(E,\ell_1)$.

Propotision 2.6.

a) $(\Sigma\oplus E_n)_1$ *is a Schur space (i.e., weakly convergent sequences are norm convergent) if and only if so is every* E_n.

b) $(\Sigma\oplus E_n)_1$ *is weakly sequentially complete if and only if so in each* E_n.

c) $(\Sigma\oplus E_n)_1$ *contains a copy of* c_0 *if and only if there exists* $n\in \mathbb{N}$ *such that* E_n *contains a copy of* c_0.

Proof. It follows from corollary 1.6(a) and the following characterizations, the first and second of which are obvious, and the third is due to Pelczynski ([7], lemma 1).

E is a Schur space if and only if $L(E,.)=DP(E,.)$.

E is weakly sequentially complete if and only if $L(E,.)=D(E,.)$.

E contains no copy of c_0 if and only if $L(E,.)=U(E,.)$.

In the case p=0 we can strengthen the above results. Recall that a Banach space E is said to be an h-space if each closed infinite dimensional subspace H of E contains a complemented subspace isomorphic to E. (see [10]), $\ell_p(I)$ $(1\leq p<\infty$ and $c_0(I)$ are examples of h-spaces for any index I.

Proposition 2.7.

a) If $(\Sigma\oplus E_n)_0$ *contains a reflexive, infinite dimensional complemented subspace, then there exists* $n\in \mathbb{N}$ *such that* E_n *contains a reflexive infinite dimensional, subspace.*

b) Let F *be an infinite dimensional h-space not containing a copy of* c_0. *The following assertions are equivalent:*

i) $(\Sigma\oplus E_n)_0$ *contains a complemented copy of* F.

ii) There exists $n\in \mathbb{N}$ *such that* E_n *contains a complemented copy of* F.

Proof.

a) Suppose H is an infinite dimensional, reflexive complemented subspace of $(\Sigma\oplus E_n)_0$. Then the projection P onto H is weakly compact. Hence, theorem 1.7 assures that P is the limit in the norm operator topology of the operators $\sum_{n=1}^{m} P_n \cdot \Pi_n$. As P is not strictly singular there has to exist an $n\in \mathbb{N}$ such that P_n is not strictly singular. Thus, there is an infinite dimensional closed subspace F of E_n such that the restriction $S=P_n|_F$ is an isomorphism. Hence F is reflexive.

b) Let P be a continuous projection onto an isomorphic copy of F. Because F does not contain a copy of c_0, P is unconditionally converging ([7] lemma 1) and then theorem 1.7 applies. Reasoning as in part (a), we get an $n\in \mathbb{N}$ and an infinite dimensional subspace $H \subset E_n$ such that the restriction $S=P_n|_H$ is an isomorphism onto its range. But as F is an h-space , S(H) contains a subspace M isomorphic to F and complemented in F. If $Q:F\longrightarrow M$ is a continuous projection, $S^{-1}\cdot Q\cdot P_n$ is a continuous projection from E onto $S^{-1}(M)$, which is isomorphic to F.

In particular, if we consider the h-space ℓ_p $(1\leq p<\infty)$, we obtain:

Corollary 2.8. *If* $1\leq p<\infty$, *the following assertions are equivalent:*

a) $(\Sigma\oplus E_n)_0$ *contains a complemented copy of* ℓ_p.

b) There exists $n\in \mathbb{N}$ *such that* E_n *contains a complemented copy of* ℓ_p.

REFERENCES

[1] Bombal, F. Two new classes of Banach spaces. To appear.
[2] Bombal, F, & Cembranos, P. (1985). Characterization of some classes of operators on spaces of vector-valued continuous functions. Math. Proc. Camb. Phil. Soc. 97, 137-146.
[3] Cembranos, P. (1982). Algunas propiedades del espacio de Banach C(k,E). Thesis. Pub. of the Universidad Complutense. Madrid.
[4] Grothendieck, A. (1953). Sur les applications linéaires faiblement compacts d'espaces du type C(k). Canad. J. of Math., 5, 129-173.
[5] Lindenstrauss, J. and Tzafriri L (1977). Classical Banach spaces I. Springer, Berlin.
[6] Pelczynski, A. (1962). Banach spaces in which every unconditionally converging operators is weakly compact. Bull. Acad. Pol. Sci. 10, 641-648.
[7] Pelczynski, A. (1965). On strictly singular and strictly cosingular operators I. Bull. Acad. Pol. Sci. 13, 31-36.
[8] Pietsch, A. (1978). Operator ideals. Berlin.
[9] Rosenthal, H.P. (1970). On injective Banach spaces and the spaces $L^{\infty}(\mu)$ for finite measures. Acta Math. 124, 205-248.
[10] Whitley, R.J. (1964). Strictly singular operators and their conjugates. Trans. Amer. Math. Soc. 113, 252-261.

Supported partially by CAYCIT grant 0338/84.

ON THE DUALITY PROBLEM FOR ENTROPY NUMBERS.

Hermann König
Univ. Kiel, Germany.

In this note we give a survey on some recent results concerning the duality problem for entroypy numbers and covering numbers.

If K_1 and K_2 are bounded absolutely convex sets in a Banach space Y with $\overset{\circ}{K}_2 \neq \emptyset$, we define the *covering number* $N(K_1,K_2)$ by

$$N(K_1,K_2) := \inf \left\{ N \in \mathbb{N} \mid \exists\ y_1,\ldots,y_N \in Y\ ,\ K_1 \subset \bigcup_{i=1}^{N} (\{y_i\} + K_2) \right\}.$$

Let X and Y be Banach spaces with closed unit balls B_X and B_Y. The *entropy numbers* of a (continuous linear) operator $u: X \longrightarrow Y$ are given by

$$e_k(u) = \inf \left\{ \varepsilon > 0 \mid N(u(B_X),\ \varepsilon B_Y) \leq 2^{k-1} \right\},\ k \in \mathbb{N}.$$

As a corollary to the polar decomposition theorem, one finds that the entropy numbers are self-dual if X and Y are Hilbert spaces, i.e. $e_k(u) = e_k(u^*)$. In general spaces this is false: take e.g. $X = \ell_1^2 = (\mathbb{R}^2, \|\cdot\|_1)$, $Y = \ell_2^2 = (\mathbb{R}^2, \|\cdot\|_2)$ and $u = id: \ell_1^2 \longrightarrow \ell_2^2$. Here $\|(x_i)\|_p = (\Sigma |x_i|^p)^{1/p}$. It is easy to see (geometrically) that

$$e_2(u:\ \ell_1^2 \longrightarrow \ell_2^2) = \sqrt{10}/4 > (2+\sqrt{2})/4 = e_2(u^*:\ \ell_2^2 \longrightarrow \ell_\infty^2\ ).$$

As of yet, it is unsettled in general, if the entropy numbers of an operator and its dual are proportional, i.e. whether

(1) $$\exists\ 1\le c<\infty\quad \forall\ k, u: X\longrightarrow Y \qquad c^{-1}e_k(u)\le e_k(u^*)\le c e_k(u)$$

holds or not. Note that

$$u \text{ is compact} \iff u^* \text{ is compact} \iff e_k(u)\longrightarrow 0 \iff e_k(u^*)\longrightarrow 0.$$

Even the more restricted and more accesible question whether

(2) $$\exists\ 1\le a, c<\infty\quad \forall k, u: X\longrightarrow Y \qquad c^{-1}e_{[ak]}(u)\le e_k(u^*)\le c e_{[k/a]}(u)$$

holds or not, is not known. (2) would imply that for any symmetric sequence space E one has

(3) $$(e_k(u))_{k\in\mathbb{N}} \in E \iff (e_k(u^*))_{k\in\mathbb{N}} \in E.$$

As shown recently by Tomczak-Jaegerman [7], (3) is true for $u: X\longrightarrow Y$ if X or Y is a Hilbert space. See also [5].

In some particular cases and examples positive partial answers are known. E.g. for identity maps $Id: \ell_p^n\longrightarrow \ell_q^n$ the strongest statement (1) is true, see Schütt [6]. The proofs for general operators all rely (in various forms) on a reduction to the Hilbert spaces case. Recall that a Banach space X if of <u>type 2</u> provided that

$$\exists\ 1\le c<\infty \quad \mathop{\forall x_1,\dots,x_m\in X}_{m\in\mathbb{N}} \quad \mathop{\text{Average}}_{\pm} \Big\| \sum_{i=1}^m \pm x_i \Big\|^2 \le c^2\cdot \sum_{i=1}^m \|x_i\|^2.$$

The best c is denoted by $T_2(X)$. Examples are L_p-spaces for $2\le p<\infty$ with $T_2(L_p)\le \sqrt{p}$.

<u>Proposition 1</u> [1] *Let* X *and* Y *be Banach spaces such that* X *and* Y^* *are of type* 2. *Then for any* $u: X\longrightarrow Y$ *inequalities* (2) *hold with* $a = \ln(18\ T_2(X)T_2(Y^*)\gamma(X)\gamma(Y^*))$, $c = \alpha[T_2(X)T_2(Y^*)]^2$. *Here* $\gamma(X)$ *is the K-convexity constant of* X *and* α *some numerical constant.*

The first step in the proof is the proof of (2) for operators of rank k. This easily reduces to the case of dim X = dim Y =k. The maximal volume ellipsoids contained in $B_X{}^*$ and B_Y, respectively, define Hilbert spaces H and K, respectively. By Gordon-Reisner [2] and Santalo's inequality

$$\left(\frac{\mathrm{vol}_n B_X{}^*}{\mathrm{vol}_n \xi}\right)^{1/n} \le \left(\frac{\mathrm{vol}_n \xi}{\mathrm{vol}_n B_X}\right)^{1/n} \le T_2(X),$$

ξ=unit ball of H. Volume arguments then show that $B_X{}^*$ is covered by L translates of the unit ball ξ where
$\log_2 L \le [\log_2(3T_2(X))k] + 1 =: [bk] + 1$. Thus

$$e_{[bk]+2}(i^{-1^*} : X^* \longrightarrow H) \le 1,$$

$i: X \longrightarrow H$ the formal identity map. A similar statement holds for $j^{*-1} : K \longrightarrow Y^*$. Together with the fact that $j^{-1}Ti^{-1}$: $H \longrightarrow K$ has self-dual entropy numbers (Hilbert space case), we find estimates of the form (2).

The reduction to rank k maps is done using the approximation numbers of u, $a_k(u) = \inf \left\{\|u-u_k\| \mid \mathrm{rank}\ u_k < k\right\}$. Carl showed that for $u: X \longrightarrow Y$ with X, Y^* of type 2, the geometric mean of the Gelfand numbers is proportional to the entropy numbers. The same statement holds for the geometric mean of the approximation numbers. The reduction to rank k maps uses that the approximation numbers of compact maps are self-dual.

In general Banach spaces X and Y, there is another positive partial result of form (2) for finite rank maps $u: X \longrightarrow Y$ and k larger than a multiple of the rank of u:

<u>Proposition 2</u> [3]. *For any* $\lambda>0$ *there is* $a = a(\lambda)$ *such that for any Banach spaces* X,Y *and any finite rank operator* $u: X \longrightarrow Y$ *and* $k \ge \lambda \cdot \mathrm{rank}\ u$ *one has*

$$e_{[ak]}(u) \le 2e_k(u^*), \quad e_{[ak]}(u^*) \le 2e_k(u).$$

This follows from the following duality result for covering numbers:

Proposition 3 [3]. *There is* $1 \le c < \infty$ *such that for all* $n \in \mathbb{N}$, *all compact, absolutely convex bodies* K_1, K_2 *in* $\mathbb{R}^n$ *and any* $\varepsilon > 0$

$$(4) \qquad c^{-1}N(K_2^0, \varepsilon K_1^0)^{1/n} \le N(K_1, \varepsilon K_2)^{1/n} \le cN(K_2^0, \varepsilon K_1^0)^{1/n}$$

Here K_1^0, K_2^0 *denote the polar bodies with respect to the euclidean norm on* $\mathbb{R}^n$.

To prove proposition 2 for rank $u = n$-dimensional spaces X and Y (to which it reduces easily), one has to apply proposition 3 to $K_1 = u(B_X)$, $K_2 = B_Y$. Then $K_1^0 = u^{*-1}(B_{X^*})$, $K_2^0 = B_{Y^*}$.

The proof of proposition 3 uses volume arguments as well and uses an inverse form of the Brunn-Minkowski inequality due to Milman [4] for the reduction to Hilbert spaces. Let D denote the euclidean unit ball in $\mathbb{R}^n$.

Inverse Brunn-Minkowski-inequality [4]. *There is* $1 \le c < \infty$ *such that for all compact, absolutely convex bodies* $K \subset \mathbb{R}^n$ *there is a linear map* $v: \mathbb{R}^n \longrightarrow \mathbb{R}^n$ *with* $\det v = 1$ *such that for any* $\varepsilon > 0$

$$[\mathrm{vol}_n(v(K) + \varepsilon D)]^{1/n} \le c[\mathrm{vol}_n(K)]^{1/n} + \alpha_n \varepsilon [\mathrm{vol}_n(D)]^{1/n}$$

with $\alpha_n \longrightarrow 1$ *for* $n \longrightarrow \infty$.

By the Brunn-Minkowski inequality, always

$$[\mathrm{vol}_n(v(K) + \varepsilon D)]^{1/n} \ge [\mathrm{vol}_n(K)]^{1/n} + \varepsilon [\mathrm{vol}_n(D)]^{1/n}$$

holds. One can use the Brunn-Minkowski inequality and its inverse to give a more precise formula for covering numbers $N(v(K), D)$ in terms of volumes. A special case of theorem 2 of [3] is

Proposition 4. *There is* $1 \le c < \infty$ *such that for all compact, absolutely convex bodies* $K \subset \mathbb{R}^n$ *there is* $v: \mathbb{R}^n \longrightarrow \mathbb{R}^n$ *with* $\det v = 1$ *such that for all* $\varepsilon > 0$

$$\max\left(\left(\frac{\mathrm{vol}_n(K)}{\mathrm{vol}_n(D)}\right)^{1/n}/\varepsilon, 1\right) \le N(v(K), \varepsilon D)^{1/n} \le$$

$$\le c \max\left(\left(\frac{\mathrm{vol}_n(K)}{\mathrm{vol}_n(D)}\right)^{1/n}/\varepsilon, 1\right).$$

Thus the volume ratio determines the covering numbers.

REFERENCES.

[1] Gordon, Y; König, H.; Schütt, C. (1987). Geometric and probabilistic estimates for entropy and approximation numbers of operators. J. Approx. Th.

[2] Gordon, Y.; Reisner, S. (1981). Some aspects of volume estimates to various parameters in Banach spaces. Proc. Workshop Banach Space Theory. Univ. Iowa, 23-53.

[3] König, H.; Milman, V. (1987). On the covering numbers of convex bodies. GAFA seminar Tel Aviv 1984/85, in: Lecture Notes in Math., Springer.

[4] Milman, V. (1986). Inégalité de Brunn-Minkowski inverse et applications à la théorie locale des espaces normés. C.R.A.S. 302 25-28.

[5] Pajor, A.; Tomczak-Jaegermann, N. (1985). Remarques sur les nombres d'entropie d'un opérateur et son transposé. C.R.A.S. 301, 743-746.

[6] Schütt, C. (1983). Entropy numbers of diagonal operators between sequence spaces. J. Approx. Th. 40, 121-128.

[7] Tomczak-Jaegermann, N. (1987). Private communication.

MIXED SUMMING NORMS AND FINITE-DIMENSIONAL LORENTZ SPACES

G.J.O. Jameson
University of Lancaster
Great Britain

INTRODUCTION

This largely expository article describes some recent results concerning the mixed summing norm $\Pi_{p,1}$ as applied to operators on ℓ_∞-spaces (especially the finite-dimensional spaces ℓ_∞^n). The whole subject of summing operators and their norms can be said to have started with a result on $\Pi_{2,1}$ - the theorem of Orlicz (1933) that the identity operators in ℓ_2 and ℓ_1 are (2,1)-summing. However, since that time the study of mixed summing norms has been somewhat neglected in favour of the elegant and powerful theory of the "unmixed" summing norms Π_p. A breakthrough has now been provided by the theorem of Pisier [7], which does for mixed summing norms what the fundamental theorem of Pietsch does for unmixed ones (see e.g. [2]). One version of Pisier's theorem states that the operator can be factorised through a Lorentz function space $L_{p,1}(\lambda)$. Such spaces were introduced in [5], and are discussed in [1] and [4]. However, it is not easy to find a really simple outline of the definition and basic properties of these spaces adapted to the (obviously simpler) finite-dimensional case, so the present paper includes a brief attempt to provide one.

It is of particular interest to compare the value of $\Pi_{2,1}$ and Π_2 for operators on ℓ_∞ or ℓ_∞^n. It is a well-known fact, underlying the famous Grothendieck inequality, that there is a constant K, independent of n, such that for all operators T from ℓ_∞^n to ℓ_1 or ℓ_2, we have $\Pi_2(T) \le K\|T\|$. This equates to saying that $\Pi_2(T) \le K'\Pi_{2,1}(T)$ for such T. It is natural to ask whether this relationship holds for all operators from ℓ_∞^n into an arbitrary Banach space Y. It was shown in [3] that this is not the case: for each n, there is an operator T_n on ℓ_∞^n such that

$$\Pi_2(T_n) \geq \frac{1}{2} (\log n)^{1/2} \, \Pi_{2,1}(T_n).$$

It was also shown that this example gives the correct order of growth: there exists C such that for all operators on ℓ_∞^n,

$$\Pi_2(T) \leq C \, (\log n)^{1/2} \, \Pi_{2,1}(T),$$

(so the ratio of Π_2 to $\Pi_{2,1}$ grows very slowly with n). We shall describe a different approach to both these results, using Lorentz spaces. In the case of the second result, this will lead to a more general statement, which is due to Montgomery-Smith (6).

Notation

We denote by [n] the set $\{1,2,\dots,n\}$ and by ℓ_∞^n the space $\mathbb{R}^n$ (or $\mathbb{C}^n$) with supremum norm. We denote the ith coordinate of an element x of $\mathbb{R}^n$ by x(i), and the ith unit vector by e_i. For an operator T from $\ell_\infty(S)$ to another normed linear space Y, the summing norms $\Pi_p(T)$ and $\Pi_{p,1}(T)$ are defined by:

$$\Pi_p(T) = \sup \left\{ \left(\sum_i \|Tx_i\|^p \right)^{1/p} : \left\| \sum_i |x_i|^p \right\|_\infty \leq 1 \right\},$$

$$\Pi_{p,1}(T) = \sup \left\{ \left(\sum_i \|Tx_i\|^p \right)^{1/p} : \left\| \sum_i |x_i| \right\|_\infty \leq 1 \right\},$$

in which all finite sequences of elements $(x_1,\dots,x_k)$ are considered.

As usual, p' denotes the conjugate index to p: $\frac{1}{p} + \frac{1}{p'} = 1$.

Pisier's theorem

Let T be an operator defined on $\ell_\infty(S)$ (for any set S). Suppose that there is a positive linear functional φ on $\ell_\infty(S)$ such that $\|\varphi\|=1$ and $\|Tx\| \leq C\, \varphi(|x|)^{1/p} \|x\|_\infty^{1/p'}$ for all x. If $\left\| \sum_i |x_i| \right\|_\infty \leq 1$, it follows that

$$\sum_i \|Tx_i\|^p \le C^p \sum_i \varphi(|x_i|) \le C^p ,$$

so that $\Pi_{p,1}(T) \le C$.

Pisier's theorem states the converse, though without exact preservation of the constant:

Theorem 1. *Let* T *be a* (p,1)*-summing operator defined on* $\ell_\infty(S)$ *for some set* S . *Then there exist* $C \le p^{1/p}\, \Pi_{p,1}(T)$ *and a positive linear functional* φ *on* $\ell_\infty(S)$ *such that* $\|\varphi\|=1$ *and*

$$\|Tx\| \le C\, \varphi(|x|)^{1/p}\, \|x\|_\infty^{1/p'}$$

for all x *in* $\ell_\infty(S)$.

For the proof, see [7], theorem 2.4 or [2], section 14.

Note that the case p=2 can be written

$$\|Tx\|^2 \le C^2\, \varphi(|x|)\, \|x\|_\infty .$$

We shall see that Pisier's theorem can be restated in terms of factorization through a Lorentz space. Before proceeding to this, we mention a few points that relate directly to the theorem in the form just stated.

For any space E, write $\alpha(E)$ for $\alpha(I_E)$, where α is a norm applying to operators.

Corollary 1.1. *For any space* E,

$$\Pi_1(E) \le 2\, \Pi_{2,1}(E)^2 \lambda(E)^2,$$

where $\lambda(E)$ *is the projection constant of* E. *If* $\lambda(E)$ *and* $\Pi_{2,1}(E)$ *are both finite, then* E *is finite-dimensional.*

Proof. Embed E in a space $\ell_\infty(S)$, and take a projection P with $\|P\|$ close to $\lambda(E)$. Note that $\pi_{2,1}(P) \le \|P\|\, \Pi_{2,1}(E)$. There is a functional φ on $\ell_\infty(S)$

such that $\|\varphi\|=1$ and

$$\|Px\|^2 \le 2\,\Pi_{2,1}(P)^2\,\varphi(|x|)\,\|x\|$$

for all x. For $x \in E$, we have $Px = x$, so that $\|x\| \le 2\,\Pi_{2,1}(P)^2\varphi(|x|)$. It is elementary that this implies that $\Pi_1(E) \le 2\,\Pi_{2,1}(P)^2$. The final statement follows from the fact that $\Pi_1(E)$ is only finite when E is finite-dimensional (see (2)).

Unlike the corresponding theorem of Pietsch for p-summing operators, Pisier's theorem does not apply to operators defined on a subspace of $\ell_\infty(S)$. This is shown by the following example.

Example 1. Let E_n be an isometric copy of ℓ_2^n in $\ell_\infty(S)$. It is elementary that $\Pi_{2,1}(E_n) = 1$ and $\Pi_1(E_n) \ge \sqrt{n}$. Suppose that φ is a functional on $\ell_\infty(S)$ such that $\|\varphi\|=1$ and $\|x\|^2 \le K^2\,\varphi(|x|)\,\|x\|$ for all $x\in E_n$ (as would be given by Pisier's theorem applied to the identity in E_n). Then $\Pi_1(E_n) \le K^2$, so $K^2 \ge \sqrt{n}$.

Next we give an example (again due to Montgomery-Smith) to show that the extra factor $p^{1/p}$ appearing in Pisier's theorem cannot simply be removed. For this, we assume the following result of Maurey (see [2], 14.4): if T is an operator on ℓ_∞^n, then $\Pi_{2,1}(T)$ can be computed by using only disjointly supported elements of ℓ_∞^n in the definition.

Example 2. Let Y be $\mathbb{R}^3$ with norm

$$\|y\|_Y = \max\Big[|y(2)| + |y(3)|,\ |y(3)|+|y(1)|,\ |y(1)|+|y(2)|\Big].$$

Let T be the identity mapping from ℓ_∞^3 to Y. Since $\|e_i+e_j\|_Y = \|e_1+e_2+e_3\|_Y = 2$, it follows easily from the result quoted that $\Pi_{2,1}(T)^2 = 5$. If $\lambda_i \ge 0$ are such that $\|Tx\|^2 \le \sum \lambda_i |x(i)|\,\|x\|$ for all x, then $\lambda_i+\lambda_j \ge 4$ for distinct i, j, hence $\lambda_1+\lambda_2+\lambda_3 \ge 6$.

i, j, hence $\lambda_1+\lambda_2+\lambda_3 \geq 6$.

Finite-dimensional Lorentz spaces.

Let λ be a measure on $[n] = \{1,2,\ldots,n\}$. Write λ_j for $\lambda\{j\}$. For $x \in \mathbb{R}^n$ and $A \subseteq [n]$, the expression $\int_A x$ means $\sum_{j\in A} \lambda_j x(j)$, and $\int x$ stands for $\int_{[n]} x$. In the usual way $L_p(\lambda)$ denotes $\mathbb{R}^n$ with the norm $\|x\|_p^\lambda = \left(\int |x|^p\right)^{1/p}$.

For $p\geq 1$, the Lorentz function space $L_{p,1}(\lambda)$ can be defined as follows. (Note: this is not the same thing as a Lorentz sequence space!) Given an element x of $\mathbb{R}^n$, arrange its terms in order of magnitude:

$$|x(j_1)| \geq \ldots \geq |x(j_n)|.$$

Write

$$x^*(k) = |x(j_k)|,$$
$$\rho_k = \lambda_{j_1}+\ldots+\lambda_{j_k},$$

and $x^*(n+1) = \rho_0 = 0$. For $0 \leq t \leq \|x\|_\infty$, let

$$d_x(t) = \lambda\{j: |x(j)|>t\}.$$

Clearly, $d_x(t) = \rho_k$ for $x^*(k+1) \leq t < x^*(k)$.

Now define

$$\|x\|_{p,1}^\lambda = \int_0^{\|x\|_\infty} d_x(t)^{1/p}dt \qquad (1)$$

$$= \sum_{k=1}^{n} \rho_k^{1/p}[x^*(k)-x^*(k+1)] \qquad (2).$$

By Abel summation, this is also equal to

$$\sum_{k=1}^{n} x^*(k)(\rho_k^{1/p} - \rho_{k-1}^{1/p}) \qquad (3).$$

If some of the $x^*(k)$ coincide, then different choices of the j_k are possible. However, it is clear from expressions (1) and (2) that this does not affect the definition. It is not quite so clear that $\|\cdot\|_{p,1}^{\lambda}$ is a norm. We return to this point below. First we list some properties that follow at once from the definition.

Proposition 2. (i) *If* $|y| \le |x|$, *then* $\|y\|_{p,1}^{\lambda} \le \|x\|_{p,1}^{\lambda}$.

(ii) *For* $A \subseteq [n]$, $\|\chi_A\|_{p,1}^{\lambda} = \lambda(A)^{1/p}$ ($= \|\chi_A\|_p^{\lambda}$).

(iii) $\|x\|_{1,1}^{\lambda} = \|x\|_1^{\lambda}$ *for all* x.

(iv) $\|x\|_{p,1}^{\lambda} \le \rho_n^{1/p}\|x\|_{\infty}$ *(note* $\rho_n = \lambda([n])$*)*.

(v) If $\lambda \le K\mu$, *then* $\|x\|_{p,1}^{\lambda} \le K^{1/p}\|x\|_{p,1}^{\mu}$ *for all* x.

Proof. Immediate.

Proposition 3. *For all* x *in* $\mathbb{R}^n$,

$$\|x\|_{p,1}^{\lambda} \le (\|x\|_1^{\lambda})^{1/p}\|x\|_{\infty}^{1/p'}.$$

Proof. Write $y_k = x^*(k) - x^*(k+1)$, so that

$$\begin{aligned} \|x\|_{p,1}^{\lambda} &= \sum_k \rho_k^{1/p} y_k \\ &= \sum_k (\rho_k y_k)^{1/p}\, y_k^{1/p'} \\ &= \left[\sum \rho_k y_k\right]^{1/p} \left[\sum y_k\right]^{1/p'} \end{aligned}$$

by Hölder's inequality. But $\sum \rho_k y_k = \|x\|_1^{\lambda}$ and $\sum y_k = x^*(1) = \|x\|_{\infty}$.

Corollary 3.1. *Let* K_n *denote the identity operator from* ℓ_{∞}^n *to* $L_{p,1}(\lambda)$. *Then* $\Pi_{p,1}(K_n) = \|K_n\| = \lambda([n])^{1/p}$. *In particular, if* $\lambda([n]) = 1$, *then* $\Pi_{p,1}(K_n) = 1$.

Proof. This follows from the easy implication in Pisier's theorem, given that $\|x\|_1^\lambda = \varphi(|x|)$, where φ is a functional on ℓ_∞^n with norm equal to $\lambda([n])$.

The next result is the key lemma paving the way for most of our further statements concerning $L_{p,1}(\lambda)$.

Proposition 4. *Let* q *be any seminorm on* $\mathbb{R}^n$ *such that if* $A \subseteq [n]$ *and* $|y| = \chi_A$, *then* $q(Y) \le \lambda(A)^{1/p}$. *Then* $q(x) \le \|x\|_{p,1}^\lambda$ *for all* x.

Proof. With notation as before, let $f_k = \operatorname{sgn} x(j_k)e_{j_k}$. Then

$$x = \sum_{k=1}^{n} x^*(k)f_k = \sum_{k=1}^{n} u_k ,$$

where

$$u_k = \left[x^*(k)-x^*(k+1)\right](f_1+\ldots+f_k).$$

By hypothesis, $q(f_1+\ldots+f_k) \le \rho_k^{1/p}$, so

$$\begin{aligned} q(x) &\le \sum_k q(u_k) \le \\ &\le \sum_k \left[x^*(k)-x^*(k+1)\right]\rho_k^{1/p} = \\ &= \|x\|_{p,1}^\lambda . \end{aligned}$$

Corollary 4.1. $\|x\|_p^\lambda \le \|x\|_{p,1}^\lambda$ *for all* x.

We mention briefly some further basic facts relating to $L_{p,1}(\lambda)$ which can be proved easily with the help of Proposition 4. Except for the fact that $\|\cdot\|_{p,1}^\lambda$ is a norm, these results are not needed for our further development.

Proposition 5. *Let* $(r_1,\ldots,r_n)$ *be any permutation of* $[n]$, *and write* $\sigma_k = \lambda_{r_1}+\ldots+\lambda_{r_k}$. *Then for any* x,

$$\sum_k |x(r_k)|\ (\sigma_k^{1/p}-\sigma_{k-1}^{1/p}) \le \|x\|^{\lambda}_{p,1}.$$

(The left hand side defines another norm, to which we apply Proposition 4)

Proposition 6. $\|\cdot\|^{\lambda}_{p,1}$ *is a norm.*

Proposition 7. *Under the duality given by* $\langle x,y\rangle = \int xy$, *the dual norm to* $\|\cdot\|^{\lambda}_{p,1}$ *is*

$$\|y\|^{\lambda}_{p',\infty} = \sup\left\{\lambda(A)^{-1/p}\int_A |y| :\ A \subseteq [n]\right\}.$$

(Apply Proposition 4 with $q(x)=|\langle x,y\rangle|$.)

Proposition 8. *Given positive elements* $x_1,\ldots,x_k$ *of* $L_{p,1}(\lambda)$, *we have*

$$\left[\sum \|x_i\|^p\right]^{1/p} \le \|\sum x_i\|,$$

where $\|\cdot\|$ *stands for* $\|\cdot\|^{\lambda}_{p,1}$. *In other words,* $L_{p,1}(\lambda)$ *satisfies a "lower p-estimate" with constant* 1.

(One proves that the dual satisfies an upper p'-estimate.)

For $p \ge q \ge 1$, one defines $\|x\|^{\lambda}_{p,q}$ to be

$$\left[\int d_x(t)^{q/p}\, dt\right]^{1/q}.$$

At the cost of some extra complexity, the above results can be adapted for this case.

The second form of Pisier's theorem

Theorem 9. *Let* T *be any operator on* ℓ_∞^n. *Then there is a probability measure* λ *on* $[n]$ *such that*

$$\|Tx\| \le p^{1/p}\Pi_{p,1}(T)\ \|x\|_{p,1}^{\lambda}$$

for all x. *Hence* $T = T_1K_n$, *where* K_n *is the identity operator from* ℓ_∞^n *to* $L_{p,1}(\lambda)$ *(so that* $\Pi_{p,1}(K_n)=1$) *and* $\|T_1\| \le p^{1/p}\ \Pi_{p,1}(T)$.

Proof. Let φ be the functional given by Theorem 1. There exist $\lambda_j \ge 0$ such that $\sum \lambda_j = 1$ and $\varphi(x) = \sum \lambda_j x(j)$ for all x. This defines a probability measure λ on $[n]$. Let $q(x) = \|Tx\|$. Apply Proposition 4 to the seminorm q. If $|y| = \chi_A$, then $q(y) \le C\varphi(\chi_A)^{1/p} = C\lambda(A)^{1/p}$, where $C \le p^{1/p}\ \Pi_{p,1}(T)$. The statement follows.

Some estimates for counting measure

Let ν denote normalised counting measure on n, so that $\nu_j = \frac{1}{n}$ for each j. Write simply $\|\cdot\|_p$ and $\|\cdot\|_{p,1}$ for the corresponding norms, so that

$$\|x\|_p = \left(\frac{1}{n}\sum |x(j)|^p\right)^{1/p},$$

$$\|x\|_{p,1} = n^{-1/p}\sum_k x^*(k)(k^{1/p}-(k-1)^{1/p}).$$

We also write $L_{p,1}^n$ for $L_{p,1}(\nu)$.

We concentrate on the case p=2. Write

$$\sum_{k=1}^{n}\frac{1}{k} = l_n ,$$

$$\sum_{k=1}^{n}\left(\sqrt{k}-\sqrt{k-1}\right)^2 = L_n.$$

Since $\sqrt{k}-\sqrt{k-1} = 1/\left(\sqrt{k}+\sqrt{k-1}\right)$, we have

$$\frac{1}{4}\, l_n \le L_n \le \frac{1}{4}\, l_n + 1.$$

Proposition 10. *For all* x *in* $\mathbb{R}^n$, *we have*

$$\|x\|_2 \le \|x\|_{2,1} \le \sqrt{L_n}\, \|x\|_2,$$

and the factor $\sqrt{L_n}$ *is attained.*

Proof. The left-hand inequality was proved in Corollary 4.1, and the right-hand one follows from Schwarz's inequality.

Define $a \in \mathbb{R}^n$ by: $a(k) = \sqrt{k}-\sqrt{k-1}$ for each k. Then $\|a\|_2 = \sqrt{L_n/n}$, while $\|a\|_{2,1} = L_n/\sqrt{n}$.

The next result is essentially the example in [3], presented in terms of Lorentz spaces.

Proposition 11. *Let* J_n *be the identity operator from* ℓ_∞^n *to* $L_{2,1}^n$. *Then* $\Pi_2(J_n) = \sqrt{L_n}$, *while* $\Pi_{2,1}(J_n) = 1$. *Hence there is no constant* C *such that* $\pi_2(T) \le C\, \pi_{2,1}(T)$ *for all operators on* ℓ_∞^n *and all* n.

Proof. Let I_n denote the identity operator from ℓ_∞^n to (normalised) ℓ_2^n. It is elementary that $\pi_2(I_n) = 1$. By Proposition 10, it follows that $\pi_2(J_n) \le \sqrt{L_n}$.

Let a_1 be the element a defined in Proposition 10, and let $a_2, \dots, a_n$ be defined by cyclic permutation of the terms of a_1. Then $\left(\sum a_i^2\right)^{1/2} = \sqrt{L_n}\, e$, so has norm $\sqrt{L_n}$ in ℓ_∞^n. Also,

$$\left(\sum \|a_i\|_{2,1}^2 \right)^{1/2} = L_n.$$

This shows that $\pi_2(J_n) \ge \sqrt{L_n}$.

It follows from Proposition 11 and Maurey's theorem (see e.g. [2], Theorem 11.4) that the cotype 2 constant of $L_{2,1}^n$ is at least $(L_n/3)^{1/4}$.

For a normed lattice X of functions, the q-concavity constant $M_q(X)$ is the least constant M such that

$$\left(\sum \|x_i\|^q \right)^{1/q} \le M \left\| \left(\sum |x_i|^q \right)^{1/q} \right\|$$

for any choice of finitely many elements x_i. It is elementary that $M_q(\ell_q) = 1$. We have:

Proposition 12. $M_2(L_{2,1}^n) = \sqrt{L_n}$.

Proof. It follows easily from Proposition 10 that the constant is not greater than $\sqrt{L_n}$. Equality is demonstrated by the elements $a_1, \ldots, a_n$ considered in Proposition 11.

This contrasts with the fact that $L_{2,1}^n$ satisfies a lower 2-estimate with constant 1 (Proposition 8). A much more elaborate example of this situation is given in [4], 1.f.19.

Furthermore, it follows from Proposition 8 and [4], 1.f.7 that for any $q > 2$, the spaces $L_{2,1}(\lambda)$ are uniformly q-concave (and hence uniformly of cotype q). Using this, Montgomery-Smith (unpublished) has shown that $\Pi_{2,1}(L_{2,1}^n) \to \infty$ as $n \to \infty$, its growth being at least of the order of $(\log \log n)^{1/2}$.

Comparison of other norms with $\Pi_{p,1}$.

We now describe the generalisation by Montgomery-Smith [6] of the theorem proved in [3]. By an "operator ideal norm" we mean a norm α applying at least to finite-rank operators satisfying $\alpha(AT) \le \|A\| \alpha(T)$ and $\alpha(TA) \le \alpha(T) \|A\|$ for all T, A.

Theorem 13. *Let* α *be an operator ideal norm, and let* J_n *be the identity operator from* ℓ_∞^n *to* $L_{p,1}^n$. *For any operator* T *from* ℓ_∞^n *into another normed linear space, we have*

$$\alpha(T) \le (2p)^{1/p} \alpha(J_{2n}) \Pi_{p,1}(T).$$

Proof. Let λ and K_n be as in Theorem 9. Let s_j be the least integer such that $\lambda_j \le s_j/n$ (also, let $s_j = 1$ if $\lambda_j = 0$). Then $\sum (s_j/n) \le 2$, so $N \le 2n$, where $N = \sum s_j$. Write $\mu_j = s_j/N$. Then $\mu_j \ge \frac{n}{N} \lambda_j \ge \frac{1}{2} \lambda_j$. This defines a measure μ on $[N]$ with $\mu \ge \frac{1}{2} \lambda$, so that the identity $I_{\mu,\lambda}$ from $L_{p,1}(\mu)$ to $L_{p,1}(\lambda)$ has norm not greater than $2^{1/p}$.

Given $x \in \mathbb{R}^n$, define $\tilde{x} \in \mathbb{R}^N$ by repeating the co-ordinate $x(j)$ s_j times. Then $\|\tilde{x}\|_{p,1}^{\nu} = \|x\|_{p,1}^{\mu}$, where ν denotes normalised counting measure on $[N]$, since the elements are equimeasurable. Let $E = \{\tilde{x}: x \in \mathbb{R}^n\}$, and write E_∞^N, $E_{p,1}^N$ for E with the norms of ℓ_∞^N, $L_{p,1}^N$ respectively. Then K_n factorises as follows;

$$\ell_\infty^n \xrightarrow{I_\infty} E_\infty^N \xrightarrow{J_N} E_{p,1}^N \xrightarrow{Q} L_{p,1}(\mu) \xrightarrow{I_{\mu,\lambda}} L_{p,1}(\lambda)$$

in which $I_\infty(x) = \tilde{x}$ and $Q(\tilde{x}) = x$. Since I_∞ and Q are isometries and $\alpha(J_N) \le$ $\le \alpha(J_{2n})$, the statement follows.

The result proved in (3) -with an improved estimate of the constant- follows at once, since $\Pi_2(J_n) = \sqrt{L_n}$ (Proposition 11):

Corollary 13.1. *For any operator* T *on* ℓ_∞^n ,

$$\Pi_2(T) \le 2 \sqrt{L_{2n}}\, \Pi_{2,1}(T).$$

It is elementary that $\Pi_{2,1}(T) \le \kappa_2(T) \le \Pi_2(T)$, where κ_2 denotes the (Rademacher) cotype 2 constant. The question was raised in (3) of whether κ_2 is equivalent to either Π_2 or $\Pi_{2,1}$ for operators on ℓ_∞^n. In

[6], Montgomery-Smith goes on to show that in fact

$$\kappa_2(T) \le c(\log \log n)\, \Pi_{2,1}(T),$$

so that κ_2 is not equivalent to Π_2. The method uses Theorem 13 (with $\propto$ equal to κ_2) and involves some delicate estimates. It is also shown that the Gaussian cotype 2 of such operators is not equivalent to the Rademacher cotype 2.

References.

[1] Bergh, J. and Löfström, J. (1976). Interpolation Spaces. Springer-Verlag.

[2] Jameson, G.J.O. (1987). Summing and Nuclear Norms in Banach Space Theory. Cambridge University Press.

[3] Jameson, G.J.O. (1987). Relations between summing norms of mappings on ℓ_∞^n. Math. Zeitschrift 194, 89-94.

[4] Lindenstrauss, J. & Tzafriri, J. (1979). Classical Banach Spaces II. Springer-Verlag.

[5] Lorentz, G.G. (1950). Some new functional spaces . Ann. of Math. 51, 37-55.

[6] Montgomery-Smith, S.J. (to appear). On the cotype of operators from ℓ_∞^n. Israel J. Math.

[7] Pisier, G. (1986). Factorization of operators through $L_{p\infty}$ or L_{p1} and non-commutative generalizations. Math. Ann. 276, 105-136.

ON THE EXTENSION OF CONTINUOUS 2-POLYNOMIALS IN NORMED LINEAR SPACES.

Carlos Benítez
Dpto de Matemáticas, Univ. de Extremadura, Badajoz, Spain.

María C. Otero
Dpto. de Análisis Matemático, Univ. de Santiago, Spain

Abstract. We give an example of a normed linear space whose norm is not induced by an inner product and such that every continuous 2-polynomial defined on a linear subspace of it can be extended to the whole space preserving the norm.

INTRODUCTION.

Let E be a normed linear space over $\mathbb{K}$ ($\mathbb{R}$ or $\mathbb{C}$). We shall call 2-polynomial on E to every mapping $P: E \longrightarrow \mathbb{K}$ such that $P(x) = \Pi(x,x)$, where Π is a bilinear functional defined on ExE.

Let us recall that the bilinear functional Π associated to P in the above sense is unique if it is symmetric and that P is cotinuous if and only if the same is Π, or if and only if it has a finite norm

$$\|P\| = \sup\left\{|P(x)|: \|x\|=1\right\} = \sup\left\{|P(x)|\cdot\|x\|^{-2}: x\neq 0\right\}$$

It is an old open problem the characterization of the normed linear spaces of dimension ≥ 3 satisfying the property that every continuous 2-polynomial defined on a linear subspace of it can be extended to the whole space preserving the norm.

In Aron and Berner (1978) appears an example, that they attribute to R.M. Schottenloher, of a normed linear space in which not every continuous 2-polynomial defined in a linear subspace of it can be extended to the whole space preserving the norm.

On the other hand, not for 2-polynomials but for bilinear functionals, it is conjectured in Hayden (1967) that the inner product spaces are the only normed linear spaces satisfying and analogous property of extension preserving the norm.

In this paper we give an example (as far as we know the first to be published) of a simple family of real normed linear spaces of dimension 3, whose norm is not induced by an inner product and such that every 2-polynomial defined in a linear subspace of it can be extended to the whole space preserving the norm.

In a certain sense our example is intermediate between the negative example of Schottenloher and the well known positive example of the inner product spaces, since in the most simple case of the real 3-dimensional spaces the unit ball of a inner product space is defined by a quadric, in the example of Schottenloher such ball is defined by the intersection of three quadrics and in our example it is defined by the intersection of two quadrics.

RESULTS.

Firstly we summarize for the real case the example of Schottenloher. With the same arguments used in Aron & Berner (1978) for the space $\mathbb{C}^3$ it is not difficult to see that in the space $E=\mathbb{R}^3$ endowed with the norm $\|(x,y,z)\|= \sup(|x|,|y|,|z|)$, the 2-polynomial defined by $P(x,y,z)=y^2+yz+z^2$ cannot be extended from the linear subspace $L=\{(x,y,z): x+y+z=0\}$ to the whole space E preserving the norm, since $\|P\|_L=1$ and for every extension $\tilde{P}$ of P to E we have that $\|\tilde{P}\|_E>1$.

In geometrical terms this means that the ellipse $y^2+yz+z^2=1$, $x+y+z=0$ is around the cube $S=\{(x,y,z):\sup(|x|,|y|,|z|)=1\}$ but such ellipse is not the section by the plane $x+y+z=0$ of any quadric circumscribed to the mentioned cube.

We shall need for our example a previous lemma

Lemma 1. *If* A *and* B *are 2-polynomial in* $\mathbb{R}^2$ *such that*

$$0 \le \sup\,[A(x),B(x)],\ (x\in \mathbb{R}^2)$$

then there exists $0\le t\le 1$ *such that*

$$0 \le tA(x)+(1-t)B(x), \quad (x\in \mathbb{R}^2).$$

Proof. It is obvious when either $A(x)$ of $B(x)$ are non negative for every $x\in \mathbb{R}^2$.

In the other case let y and z be such that

$$A(y) < 0 \le B(y), \; B(z)<0\le A(z)$$

Since $(A-B)(y)$ and $(A-B)(z)$ have opposite signs there exists a point u in the segment that joints y and z, and a point v in the segment that joints y and -z such that

$$0 \le A(u) = B(u), \quad 0 \le A(v)=B(v)$$

Let α and β be the symmetric bilinear functionals associated to A and B. Expanding $A(ru+sv)$ and $B(ru+sv)$ for $r,s\in \mathbb{R}$, it is easy to see that

$$\alpha(u,v) < 0, \; \beta(u,v) >0$$

Then from the expansion of $[tA+(1-t)B](ru+sv)$ we obtain that the wanted t is such that

$$t\alpha(u,v) + (1-t)\beta(u,v) = 0.$$

Corollary. *If* P, A, B *are 2-polynomials in* $\mathbb{R}^2$ *such that*

$$P(x) \le \sup\,[A(x),B(x)], \quad (x\in\mathbb{R}^2)$$

then there exists $0\le t\le 1$ *such that*

$$P(x)\le tA(x)+(1-t)B(x), \quad (x\in \mathbb{R}^2).$$

Proposition 1. *Let* E *be the linear space* $\mathbb{R}^3$ *endowed with a norm whose unit sphere is given by*

$$S = \{x\in \mathbb{R}^3 : \sup[A(x),B(x)]=1\}$$

where A and B are positive semidefinite 2-polynomials. Then every 2-polynomial defined in a linear subspace of E can be extended to E preserving the norm.

Proof. We shall consider only the non trivial case in which P is a non zero 2-polynomial defined in a two dimensional linear subspace $L = \ker \phi$, where ϕ is a non zero linear functional in E. We shall distinguish between the cases P positive semidefinite (the case negative is analogous) and P indefinite.

Let P be positive semidefinite. Since

$$P(y) \le \|P\|_L \sup [A(y),B(y)] \ , \ (y\in L)$$

it follows from the above corollary the existence of $0\le t\le 1$ such that

$$P(y) \le \|P\|_L \ [A(y)+(1-t)B(y)], \ (y\in L)$$

Let γ be the symmetric bilinear functional on ExE associated to tA+(1-t)B. Since γ is positive semidefinite in ExE and non zero in LxL there exists $u\in E\setminus L$ such that $\gamma(u,y)=0$ for every $y\in L$. Then the linear funtional $\gamma(u,.)$ is either zero or proportional to ϕ. In both cases it suffices to define

$$\tilde{P}(x) = P\left[x - \frac{\phi(x)}{\phi(u)} u\right], \ (x\in E)$$

to obtain

$$|\tilde{P}(x)| = \tilde{P}(x) \le \|P\|_L [tA+(1-t)B] \left[x - \frac{\phi(x)}{\phi(u)} u \right]$$
$$\le \|P\|_L \ [ta+(1-t)B](x) \le \|P\|_L \sup [A(x),B(x)], \ (x\in E)$$

as we wish to prove.

Now let P be indefinite, Π be the symmetric bilinear functional associated to P and $y_1, y_2 \varepsilon$ L be such that

$$P(y_1) = \sup\{P(y):\ y \in L, \|y\|\le 1\},\ P(y_2) = \inf\{P(y): y \in L,\ \|y\|\le 1\}$$

Then $y_1, y_2 \in S$, $P(y_1)>0$, $P(y_2)<0$ and

$$P(y_i)P(y) \leq \Pi^2(y_i,y) \leq P^2(y_i) \sup[A(y),B(y)], \quad (y \in L), \quad (i=1,2)$$

where the first inequality follows from the fact that P is indefinite and the second is true since in other case there would exist $y_0 \in S \cap L$ such that $\Pi^2(y_i,y_o) > P^2(y_i)$ and hence

$$[P(ty_0+(1-t)y_i)-P(y_i)][P(-ty_0+(1-t)y_i)-P(y_i)] < 0$$

for t>0 sufficiently small, in contradiction with the definition of y_i.

Let ψ_1 and ψ_2 be norm-preserving extensions to E of the non zero linear functionals $\Pi(y_1,.)$ and $\Pi(y_2,.)$ such that ψ_1 is proportional to ψ_2 when so are $\Pi(y_1,.)$ and $\Pi(y_2,.)$. Then

$$\psi_i^2(x) \leq P^2(y_i) \sup [A(x),B(x)], \quad (i=1,2), \quad (x \in E)$$

and if we take $v \in (\ker \psi_1 \cap \ker \psi_2)\backslash L$ and define

$$\tilde{P}(x) = P\left[x - \frac{\phi(x)}{\phi(v)} v\right], \quad (x \in E)$$

we obtain

$$P(y_i)\tilde{P}(x)=P(y)P\left[x - \frac{\phi(x)}{\phi(v)} v\right] \leq \Pi^2\left[y_i, x - \frac{\phi(x)}{\phi(v)} v\right] =$$

$$= \psi_i^2\left[x - \frac{\phi(x)}{\phi(v)} v\right] = \psi_i^2(x) \leq$$

$$\leq P^2(y_i) \sup [A(x),B(x)], \quad (x \in E)$$

from which it follows that

$$|\tilde{P}(x)| \leq \sup [P(y_1), |P(y_2)|] \sup [A(x),B(x)] =$$

$$= \|P\|_L \sup [A(x),B(x)], \quad (x \in E).$$

as we wish to prove.

SOME REMARKS.

A classical and involved theorem due to Blaschke (1916) and Kakutani (1939), says that the unit sphere S of a norm in $\mathbb{R}^3$ is an ellipsoid if and only if every section of S by an homogeneous plane can be extended to a cylinder supporting (or circumscribed to)S.

On the other hand the non less classical theorem of Hahn and Banach is a generalization or abstraction of the obvious fact that every straight line supporting S (obviously in an homogeneous plane) can be extended to a plane also supporting S.

Roughly speaking we have treated in this paper with an intermediate problem which has an intermediate answer. Namely the hypothetic extension of conics supporting S in an homogeneous plane to quadrics also sopporting S.

With regard to this problem we have seen that the above extension is not always possible when S is a cube (Schottenloher example) and it is always possible when S is a "barrel" (our example)

From this stimulating points of view (a little far, certainly, from the barreled spaces) we can state in easy terms many interesting questions about this old and seemingly difficult problem.

Are the barrels the only three dimensional examples for which is valid a quadratic Hahn-Banach theorem?.

What about cylindrical extensions of circumscribed conics?. What about either dimension >3, or 2n-polynomials, or non homogeneous polynomials?. What about extension of bilinear functionals?. Etc, etc.

THE COMPLEX CASE.

As we have pointed out before the Schottenloher example was given originally in the complex space $\mathbb{C}^3$. We translated it automatically to the space $\mathbb{R}^3$ and we gave our example in this real space.

In fact, as a corollary of the following elementary lemma we obtain that the problem of extension preserving the norm of continuous 2-polynomials in complex spaces, can be reduced in all the cases to the real case.

Lemma 2. *Let* E *be a complex normed linear space,* $P: E \longrightarrow \mathbb{C}$ *be a continuous complex 2-polynomial in* E *and* Π_1 *be the continuous symmetric real bilinear functional associated to the continuous real 2-polynomial* $P_1 = \mathrm{Re}\ P$.

Then $P(x) = \Pi_1(x,x) - i\Pi_1(ix,x)$, $(x \in X)$ *and* $\|P\| = \|P\|_1$.

Proof. Let

$$P(x) = P_1(x) + iP_2(x), \text{ with } P_1(x),\ P_2(x) \in \mathbb{R}$$

and let Π (resp. Π_1, Π_2) be the continuous symmetric complex (resp. real) bilinear functional associated to P (resp. P_1, P_2).

The first part of the lemma follows from the equalities

$$\Pi_1(ix,x) + i\Pi_2(ix,x) = \Pi(ix,x) = i\Pi(x,x) = -\Pi_2(x,x) + i\Pi_1(x,x)$$

The second part is an inmediate consequence of the fact that for every $x \in E$ there exists $\vartheta_x \in \mathbb{C}$ such that $|\vartheta_x| = 1$ and $P(\vartheta_x x) \in \mathbb{R}$.

Proposition 2. *Let* E *be a complex normed linear space,* L *a linear subspace of* E *and* $P: L \longrightarrow \mathbb{C}$ *a complex continuous 2-polynomial. If the continuous real 2-polynomial* $P_1 = \mathrm{Re}\ P$ *can be extended to a real 2-polynomial in* E *of the same norm, then the same is valid for the original complex 2-polynomial* P.

Proof. If

$$\tilde{P}_1 = \Pi_1(x,x) \ , \ (x \in E)$$

is the extension of P_1 to E, then

$$\tilde{P}(x) = \Pi_1(x,x) - i\Pi_1(ix,x), \ (x \in X)$$

is the extension of P to E.

REFERENCES.

[1] Aron, R.M. & Berner, P.D. (1978). A Hahn-Banach extension theorem for analytic mappings. Bull. Soc. Math. Franc. 106, 3-24.
[2] Benitez C. & Otero, M.C. (1986). Approximation of convex sets by quadratic sets and extension of continuous 2-polynomials. Boll. Un. Mat. Ital. , Ser. VI, Vol. V-C N.1, 233-243.
[3] Blaschke, W. (1916). Kreis un Kugel. (1956). Reprinted Walter de Gruyter Berlin.
[4] Hayden, T.L. (1967a). The extension of bilinear functionals. Pacific J. Math. 22, 99-109.
[5] Hayden, T.L. (1967b). A conjecture on the extension of bilinear functionals. Amer. Math. Monthly, 1108-1109.
[6] Kakutani, S. (1939). Some Characterizations of Euclidean space. Japan Jour. Math. 16, 93-97.
[7] Marinescu, G. (1949). The extension of bilinear functionals in general euclidean spaces. Acad. Rep. Pop. Romane. Bull. Sti. A. 1, 681-686.
[8] Moraes, L.A. (1984). The Hahn-Banach extension theorem for some spaces of n-homogeneous polynomials. Functional Analysis: Surveys and Recent Results III. North-Holland, 265-274.
[9] Nachbin,L. (1969). Topologies on Spaces of Holomorphic Functions. Springer Verlag.

ON SOME OPERATOR IDEALS DEFINED BY APPROXIMATION NUMBERS

Fernando Cobos
Dpto. de Matemáticas, Univ. Autónoma de Madrid, Madrid, Spain

Ivam Resina
Inst. de Matemática, Univ. Estadual de Campinas, S. Paulo, Brasil

To Professor Antonio Plans.

Abstract. We prove a representation theorem in terms of finite rank operators for operators belonging to $\mathcal{L}_{\infty,\infty,\gamma}$. Some information on the tensor product of operators belonging to these ideals is also obtained.

INTRODUCTION.

The n-th approximation number $a_n(T)$ of a bounded linear operator $T \in \mathcal{L}(E,F)$ acting between the Banach spaces E and F, is defined as

$$a_n(T) = \inf\left\{ \|T-T_n\| : T_n \in \mathcal{L}(E,F),\ \text{rank } T_n < n \right\},\quad n=1,2,\ldots$$

(see [3], [5])

For $0 < \gamma < \infty$ the ideal $\left[\mathcal{L}_{\infty,\infty,\gamma}, \sigma_{\infty,\infty,\gamma}\right]$ is formed by all operators T between Banach spaces, with a finite quasi-norm

$$\sigma_{\infty,\infty,\gamma}(T) = \sup\{(1+\log n)^{\gamma}\, a_n(T) : n=1,2,\ldots\}.$$

Weyl ideals $\mathcal{L}^{(x)}_{\infty,\infty,\gamma}$ are defined in a similar way, by substituting approximation numbers for Weyl numbers $(x_n(T))$. Ideals $\mathcal{L}^{(x)}_{\infty,\infty,\gamma}$ have been studied by the authors in [1]. Since $x_n(T) \le a_n(T)$ for all $n \in \mathbb{N}$ (see [5]), as a direct consequence of [1], Thm. 3, we have

Theorem 1. *Let* $0 < \gamma < \infty$. *Then there is a constant* $M = M(\gamma)$ *such that for any complex Banach space* E *and any operator* $T \in \mathcal{L}_{\infty,\infty,\gamma}(E,E)$ *the following holds*

$$\sup \{ (1+\log n)^{\gamma} |\lambda_n(T)| : n\in \mathbb{N} \} \leq M \sigma_{\infty,\infty,\gamma}(T).$$

Here $(\lambda_n(T))$ denotes the sequence of all eigenvalues of the compact operator T counted accoding to their algebraic multiplicities and ordered such that $|\lambda_1(T)| \geq |\lambda_2(T)| \geq \dots \geq 0$.

In this note we continue the study of $\mathcal{L}_{\infty,\infty,\gamma}$-ideals. We derive a representation theorem for the elements of $\mathcal{L}_{\infty,\infty,\gamma}$ in terms of finite rank operators. This result is on the same lines as we established in [2] for the case of the ideals $\mathcal{L}_{\infty,q,\gamma}$ $(0 < q < \infty,\ -1/q < \gamma < \infty)$. We also obtain some information on the tensor porduct of operators belonging to the scale of the ideals $\mathcal{L}_{\infty,\infty,\gamma}$.

MAIN RESULTS.

Theorem 2. *Let* $0 < \gamma < \infty$ *and* E, F *be Banach spaces. Then the following are equivalent for* $T \in \mathcal{L}(E,F)$:

(1) $T \in \mathcal{L}_{\infty,\infty,\gamma}(E,F)$.

(2) *There are operators* $T_n \in \mathcal{L}(E,F)$ *of rank* $T_n \leq 2^{(2^n)}$ *such that* $T = \sum_{n=0}^{\infty} T_n$ *converges in the operator norm and*

$$\sup \left\{ 2^{n\gamma} \|T_n\| : n=0,1,\dots \right\} < \infty$$

In this case

$$\sigma^{*}_{\infty,\infty,\gamma}(T) = \inf \left\{ \sup\left\{ 2^{n\gamma} \|T_n\| : n=0,1,\dots \right\} : T = \sum_{n=0}^{\infty} T_n , \ \operatorname{rank} T_n \leq 2^{(2^n)} \right\}$$

defines a quasi-norm equivalent to $\sigma_{\infty,\infty,\gamma}$.

Proof. Put $\nu_n = 2^{(2^n)}$, $n = 0,1,\ldots$, and let $T \in \mathcal{L}_{\infty,\infty,\gamma}(E,F)$. We can choose $L_n \in \mathcal{L}(E,F)$ such that

$$\operatorname{rank} L_n < \nu_n \quad \text{and} \quad \|T-L_n\| \le 2a_{\nu_n}(T).$$

Write

$$T_0 = 0,\ T_1 = L_0 \text{ and } T_n = L_{n-1} - L_{n-2} \quad \text{for } n = 2,3,\ldots$$

We have

$$\left\| T - \sum_{j=0}^{n+1} T_j \right\| = \|T - L_n\| \le 2a_{\nu_n}(T) \longrightarrow 0 \text{ as } n \longrightarrow \infty.$$

Furthermore

$$\operatorname{rank} T_n < \nu_{n-1} + \nu_{n-2} < \nu_n, \quad \|T_1\| \le 3a_1(T) \qquad \text{and}$$

$$\|T_n\| \le \|L_{n-1} - T\| + \|T - L_{n-2}\| \le 4a_{\nu_{n-2}}(T), \quad n = 2,3,\ldots$$

Whence the monotonicity of approximation numbers yields

$$\begin{aligned}
&\sup\left\{ 2^{n\gamma} \|T_n\| : n = 0,1,\ldots \right\} \\
&= \sup\left\{ 2^{\gamma} \|T_1\|,\ 2^{(n+2)\gamma} \|T_{n+2}\| : n = 0,1,\ldots \right\} \\
&\le 2^{3\gamma+2} \sup\left\{ a_1(T),\ a_2(T),\ 2^{(n-1)\gamma} a_{\nu_n}(T) : n = 1,2,\ldots \right\} \\
&\le c_1 \sup\Big\{ a_1(T),\ (1 + \log 2)^{\gamma} a_2(T), \\
&\qquad \max\left\{ (1 + \log k)^{\gamma} a_k(T) : \nu_{n-1} < k \le \nu_n \right\} : n=1,2,\ldots \Big\} \\
&= c_1 \sup\left\{ (1 + \log n)^{\gamma} a_n(T) : n = 1,2,\ldots \right\}
\end{aligned}$$

where the constant c_1 depends only on γ.

Therefore

$$\sigma^*_{\infty,\infty,\gamma}(T) \le c_1 \sigma_{\infty,\infty,\gamma}(T).$$

Assume conversely that T admits a representation $T = \sum_{n=0}^{\infty} T_n$ as in (2). Since

$$\operatorname{rank} \sum_{k=0}^{n-1} T_k \le \sum_{k=0}^{n-1} \nu_k < \nu_n \ , \ n = 1,2,\dots$$

it follows that

$$a_{\nu_n}(T) \le \left\| T - \sum_{k=0}^{n-1} T_k \right\| \le \sum_{k=n}^{\infty} \|T_k\| = \sum_{k=n}^{\infty} (2^{-k\gamma} 2^{k\gamma} \|T_k\|)$$

$$\le 2^{-n\gamma} (1 - 2^{-\gamma})^{-1} \sup \left\{ 2^{k\gamma} \|T_k\| : \ k = n, \ n+1, \dots \right\}$$

Analogously

$$a_{\nu_0}(T) \le a_1(T) \le \sum_{k=0}^{\infty} \|T_k\|$$

$$\le (1 - 2^{-\gamma})^{-1} \sup \left\{ 2^{k\gamma} \|T_k\| \ : \ k = 0,1,\dots \right\} .$$

Using again the monotonicity of approximation numbers, we obtain

$$\sigma_{\infty,\infty,\gamma}(T) =$$

$$= \sup \left\{ a_1(T), \max \left\{ (1 + \log k)^{\gamma} a_k(T) : \ \nu_n \le k < \nu_{n+1} \right\} : n = 0,1,. \right\}$$

$$\le c_2 \sup \left\{ a_1(T), \ 2^{n\gamma} a_{\nu_n}(T) : \ n = 0,1,\dots \right\}$$

$$\le c_3 \sup \left\{ 2^{n\gamma} \|T_n\| \ : \ n = 0,1,\dots \right\} .$$

The constants c_2 and c_3 only depend on γ. This completes the proof. □

In order to derive a result on the tensor product of operators belonging to these ideals, we now recall some elementary facts concerning the theory of tensor products (see, for example, [4]).

By a cross-norm τ we mean a norm which is simultaneously defined on all algebraic tensor product $E \otimes F$ of Banach spaces E and F, such that

$$\tau\,(x \otimes y) = \|x\|\ \|y\| \text{ for all } x \in E,\ y \in F.$$

The complete hull of $E \otimes F$ with respect to the topology generated by τ is denoted by $E \hat{\otimes}_\tau F$.

The algebraic tensor product $S \otimes T$ of two continuously linear operators $S \in \mathcal{L}(E,E_0)$ and $T \in \mathcal{L}(F,F_0)$ is the linear operator from $E \otimes F$ into $E_0 \otimes F_0$ defined uniquely by

$$(S \otimes T)\left(\sum_{j=1}^{n} x_j \otimes y_j\right) = \sum_{j=1}^{n} Sx_j \otimes Ty_j\ ,\ x_j \in E,\ y_j \in F.$$

A croos-norm is called a tensor norm provided that for all such maps the following holds

$$\tau\left(\sum_{j=1}^{n} Sx_j \otimes Ty_j\right) \le \|S\|\ \|T\|\ \tau\left(\sum_{j=1}^{n} x_j \otimes y_j\right).$$

In this case $S \otimes T$ admits a unique τ-continuous extension acting from $E \hat{\otimes}_\tau F$ into $E_0 \hat{\otimes}_\tau F_0$ which is denoted by $S \hat{\otimes}_\tau T$.

Theorem 3. *Let* $0 < \beta < \gamma < \infty$ *and let* τ *be a tensor norm. Then* $S \in \mathcal{L}_{\infty,\infty,\beta}(E,E_0)$ *and* $T \in \mathcal{L}_{\infty,\infty,\gamma}(F,F_0)$ *imply*

$$S \hat{\otimes}_\tau T \in \mathcal{L}_{\infty,\infty,\beta}(E \hat{\otimes}_\tau F,\ E_0 \hat{\otimes}_\tau F_0).$$

Proof. Consider representations

$$S = \sum_{k=0}^{\infty} S_k \ , \ \operatorname{rank} S_k \le 2^{(2^k)} = \nu_k \ , \ \sup \left\{ 2^{k\beta} \ \|S_k\| \ : \ k = 0,1,\ldots \right\} < \infty \quad \text{and}$$

$$T = \sum_{\ell=0}^{\infty} T_\ell \ , \ \operatorname{rank} T_\ell \le \nu_\ell \ , \ \sup \left\{ 2^{\ell\gamma} \ \|T_\ell\| : \ \ell = 0,1,\ldots \right\} < \infty.$$

Put $R_0 = R_1 = 0$ and $\qquad R_n = \sum_{k+\ell=n-2} S_k \hat{\otimes}_\tau T_\ell \ , \ n = 2,3,\ldots$

We have

$$S \hat{\otimes}_\tau T = \sum_{n=0}^{\infty} R_n \text{ with } \operatorname{rank} R_n \le \sum_{k+\ell=n-2} \nu_k \nu_\ell \le \nu_{n-2} \sum_{k=0}^{n-2} \nu_k < \nu_{n-2} \, 2 \, \nu_{n-2} < \nu_n .$$

Moreover, it follows from

$$2^{n\beta} \ \|R_{n+2}\| \le 2^{n\beta} \sum_{k+\ell=n} \|S_k\| \ \|T_\ell\| =$$

$$= \sum_{k+\ell=n} 2^{k\beta} \ \|S_k\| \ 2^{\ell\gamma} \ \|T_\ell\| \ 2^{\ell(\beta-\gamma)}$$

$$\le \sup_{k\ge 0} \left\{ 2^{k\beta} \ \|S_k\| \right\} \ \sup_{\ell\ge 0} \left\{ 2^{\ell\gamma} \ \|T_\ell\| \right\} \sum_{j=0}^{\infty} 2^{j(\beta-\gamma)}$$

that $\qquad \sup \left\{ 2^{n\beta} \ \|R_n\| \ : \ n = 0,1,\ldots \right\} < \infty.$

Consequently, Theorem 2 implies

$$S \hat{\otimes}_\tau T \in \mathcal{L}_{\infty,\infty,\beta}(E \hat{\otimes}_\tau F, \ E_0 \hat{\otimes}_\tau F_0). \qquad \square$$

The stability under tensor norms of the operator ideals $\mathcal{L}_{\infty,q,\gamma}$ $(0 < q < \infty \ , \ -1/q < \gamma < \infty)$ has been studied by the authors in [2].

REFERENCES.

[1] Cobos, F. & Resina, I. Some operator ideals related to γ-summing operators. Preprint.

[2] Cobos, F. & Resina, I. Representation theorems for some operator ideals. J. London Math. Soc., to appear.

[3] König, H. (1986). Eigenvalue distribution of compact operators, Birhäuser, Baswl.

[4] Köthe, G. (1969/79). Topological vector spaces. Vol. I and II. Springer, New York-Heidelberg-Berlin.

[5] Pietsch, A. (1987). Eigenvalues and s-numbers, Cambridge Univ. Press, Cambridge.

The first author was supported in part by M.E.C. Programa de Cooperacion con Iberoamerica, and the second by FAPESP-BRASIL (Proc. 86-0964-0).

SOME REMARKS ON THE COMPACT NON-NUCLEAR OPERATOR PROBLEM

Kamil John (Praha)
Math. Inst. of Czechoslovak Academy of Sciences, Praha.

Let us consider the following long standing question [2,3]:

(*) Suppose that X,Y are infinite dimensional Banach spaces. Does then exist a compact non-nuclear operator $T: X \longrightarrow Y$?

The problem has positive answer under quite general assumptions [1,10,15]. Positive answer of the finite-dimensional version of (*) was given in [12]. Close to (*) is another problem posed by Grothendieck and solved by the example of Pisier's space P [12]:
P is an infinite dimensional Banach space such that

1) $P \otimes_{\varepsilon} P = P \otimes_{\pi} P$

and

2) P and P' are of cotype 2.

G. Pisier already observed [13] that if P were reflexive or merely if P^* had R.N. property then every bounded operator $T: P \longrightarrow P'$ would be nuclear and thus (*) would have negative answer. Indeed, 1) implies that every $T: P \longrightarrow P'$ is integral. Unfortunately no reflexive Pisier's space P is known to exist. We may also translate the problem into the category of locally convex spaces:

Suppose that X,Y are non-nuclear locally convex spaces. Does then exist a compact non-nuclear operator $T: X \longrightarrow Y$? This generalized problem has a negative answer, [6,7]: There is a non-nuclear F space X and a non-nuclear DF space Z = Y' such that every continuous operator $T: X \longrightarrow Z$ is (strongly) nuclear. X and Z are obtained via Pisier's observation mentioned above from an example with properties similar to 1):

There are non-nuclear F spaces X, Y such that $X \otimes_\varepsilon Y = X \otimes_\pi Y$. Moreover X, Y are hilbertizable. Due to the last fact we may conclude that integral operators $T\colon X \longrightarrow Y'$ are nuclear (c.f. [11] for details).

We feel therefore that (*) has (similarly as in the locally convex case) negative answer and that the Pisier's space P and its dual P' (possibly with some modifications) might be counterexamples to (*). In this note we observe some simple facts and special cases which support this conjecture.

First let us denote by $\mathcal{L}$, $\mathcal{K}$, $\mathcal{P}_p$, $\mathcal{N}$, $\mathcal{Y}$ respectively the classes of continuous, compact, absolutely p-summable, nuclear and integral operators. Further we will denote by $\mathcal{F}$ the set of finite rank operators, by $\mathcal{A}$ the approximable operators (i.e. $\mathcal{A} = \overline{\mathcal{F}} \subset \mathcal{L}$) and $\mathcal{D}_2 = \mathcal{P}_2^{dual} \cdot \mathcal{P}_2$ $(= \Gamma_2^*)$, where Γ_2 are operators factorizable through a Hilbert space (see [5,11,14] for details and further notation). We will say that a Banach space X is a Grothendieck's theorem space (G.T. space) if

$$\mathcal{L}(X,\ell_2) = \mathcal{P}_1(X,\ell_2).$$

More generally we say [4] that X is a Hilbert-Schmidt space (H.S.) if

$$\mathcal{L}(X,\ell_2) = \mathcal{P}_2(X,\ell_2).$$

This is equivalent to the fact that Id_X is absolutely (2,2,2) summing and thus is a seldfdual property:

i) X is H.S. $\Longleftrightarrow$ X' is H.S.

Evidently X is G.T. space implies that X is H.S. space. Conversely we have:

ii) If X is of cotype 2 then X is H.S. iff X is G.T. space [14, Theorem 5.16].

It can be shown that the property 1) of the Pisiser's space implies that P and P' are of every cotype $q > 2$ [12,9]. We do not know if 1) necessarily implies that P and P' are of cotype 2.

It is easy to see that 1) implies that P is H.S. space [8]. From this and from i) and ii) we may conclude:

Properties 1) and 2) imply that all the spaces P, P', P",... are G.T. spaces.

Also we have: Let $\mathcal{L}(X,Y) = \mathcal{N}(X,Y)$ for some infinite dimensional Banach space Y. Then (by Dvoretzky's theorem) X verifies G.T. (c.f. [12], p. 42). In [13] Pisier (1983) proved that $\mathcal{N}(P,P)$ coincides with the approximable operators, $\mathcal{A}(P,P)$. The following observation is a straightforward generalization of this fact:

Lemma 1. *Let* X *be any of the spaces* $P^{(n)}$ (n=0,1,2,...,n-th dual of Pisiser's space $P = P^{(0)}$) *and let* Y *be any of the spaces* $P^{(m)}$ (m=0,1,2,...). *Then*

3) $\mathcal{A}(X,Y) = \mathcal{N}(X,Y)$,

or equivalently:

4) *There is* $c > 0$ *such that* $N(T) \leq c\|T\|$ *for any* $T \in \mathcal{F}(X,Y)$. (N *denotes the nuclear norm).*

Proof. (c.f. [13.14]). According to an observation made above X and Y are H.S. spaces. Of course X' and Y are of cotype 2. By Pisier's factorization theorem [14, Theorem 4.1] there exists a constant $c_1 > 0$ such that

$$\gamma_2(T) \leq c_1\|T\| \text{ for any } T \in \mathcal{F}(X,Y),$$

i.e.

$$T = BA \quad \text{with} \quad \|A\|\cdot\|B\| \leq c_1(1+\varepsilon)\ \|T\|,$$

where $A: X \longrightarrow \ell_2$, $B: \ell_2 \longrightarrow Y$.

The Hilbert-Schmidt property of X and Y' yield now $P_2(A) \leq c_2\|A\|$ and $P_2(B') \leq c_3\|B\|$, where P_2 is the absolutely 2-summing norm and c_2, c_3 are constants depending only on P. Thus we have $P_2(A)P_2(B') \leq c_1c_2c_3(1+\varepsilon)\|T\|$. This means that for $\gamma_2^*(T)$ (which is by definition the infimum of all numbers $P_2(A)P_2(B')$ in the factorizations $T = BA$ through ℓ_2) we have $\gamma_2^*(T) \le c_4\|T\|$. Now dual version [14, Theorem 4.9] of Pisier's factorization theorem yields

$$N(T) \le c_5\gamma_2^*(T) ,$$

which gives 4).

To show 3) we suppose that $\|T-T_n\| \to 0$ with some $T_n \in \mathcal{F}$. Then, by 4) $\{T_n\}$ is a Cauchy sequence in $\mathcal{N}(X,Y)$, q.e.d.

Let X,Y be as above. The problem whether $\mathcal{K}(X,Y) = \mathcal{N}(X,Y)$ reduces to the question whether $\mathcal{K}(X,Y) = \mathcal{A}(X,Y)$. This would be so if e.g. we were able to prove the factorization $\mathcal{K}(X,Y) = \Gamma_2 \cdot \mathcal{K}(X,Y)$.

Indeed, the metric approximation property of Hilbert space gives that $\Gamma_2 \cdot \mathcal{K} \subset \mathcal{A}$. We will now produce such a factorization for certain operators. Let X be any Banach space. An operator $T: X \longrightarrow X'$ is said to be positive, resp. symmetric if $\langle Tx,x\rangle \ge 0$ for all $x \in X$, resp. $\langle Tx,y\rangle = \langle x,Ty\rangle$ for all $x,y \in X$, i.e. if $T'/X=T$ (see [16]).

Lemma 2. [16, p. 123]. *Let* $T: X \longrightarrow X'$ *be positive symmetric operator. Then* $T = A'A$, *where* $A \in \mathcal{L}(X,\ell_2)$.

Lemma 3. *Let* $T = A'A$ *be as above and let* B_X *be the unit ball of X. Then on* $AB_X \subset \ell_2$ *coincide the Hilbert norm topology and the topology induced by the operator* $A': \ell_2 \longrightarrow X'$.

Proof.- [16, p. 124]. Observe that if $x_n, x \in B_X$ then

$$\|Ax_n - Ax\|^2 = (Ax_n - Ax,\ Ax_n - Ax) = \langle x_n - x,\ A'A(x_n - x)\rangle \le$$
$$\le 2\ \|A'(Ax_n - Ax)\|\ \le 2\ \|A\| \cdot \|Ax_n - Ax\|.$$

Corollary. $T = A'A$ *is compact iff* A *is compact.*

Proposition. *Every compact positive symmetric operator* $T: P \longrightarrow P'$ (*or* $T: P' \longrightarrow P$) *is nuclear.*

Proof.' Let $T \in \mathcal{K}(P,P')$. Lemma 2 and the above Corollary yield that $T = A'A$

where $A:X\longrightarrow\ell_2$ is compact. The metric approximation property of ℓ_2 implies that $A\in\mathcal{A}$. Therefore $T\in\mathcal{A}(P,P')=\mathcal{N}(P,P')$ by Lemma 1.

Acknowledgements. I am indebted to Hans Jarchow for helpful conversation on the subjet.

REFERENCES.

[1] Davies, W.; Johnson, W.B. (1974). Compact, Non-Nuclear Operators. Studia Math. 51 , 81-85

[2] Grothendieck, A.(1955). Produits tensoriels topologiques et espaces nucléaires. Memoirs AMS 16.

[3] Grothendieck, A. (1956). Résumé de la théorie métrique des produits tensoriels topologiques. Bol. Soc. Matem. Sao Paulo 8, 1-79.

[4] Jarchow, H. (1982). On Hilbert-Schmnidt spaces. Rend. Circ. Mat. Palermo (Suppl) II (2), 153-160.

[5] Jarchow, H. (1981). Locally Convex spaces. Teubner, Stutgart.

[6] John, K. (1983). Counterexample to a conjecture of Grothendieck. Math. Ann. 265, 169-179.

[7] John, K. (1986). Nuclearity and tensor products . DOGA Tr. J. Math. 10, 125-135.

[8] John, K. (1983). Tensor products and nuclearity. Lecture Notes in Math. 991, 124-129.

[9] John, K. (1981). On tensor product characterization of nuclear spaces. Math. Ann. 257, 341-353.

[10] Johnson, W.B.; König, H.; Maurey, B.; Retherford, R. (1979). Eigenvalues of p-summing and ℓ_p-type operators in Banach spaces. J. Funct. Anal. 32, 353-380.

[11] Pietsch, A. (1978). Operator ideals. Berlin.

[12] Pisier, G. (1980). Un théorème sur les opérateurs linéaires entre espaces de Banach qui se factorisent par un space de Hilbert. Ann. scient. Ec. Norm. Sup. 13, 23-43.

[13] Pisier, G. (1983). Counterexamples to a conjecture of Grothendieck. Acta Math. 151, 181-208.

[14] Pisier, G. (1986). Factorization of linear operators and geometry of Banach spaces. Amer. Math. Soc., CBMS Regional. Conf. Series in Math., No. 60.

[15] Tseitlin, I.I. (1973).On a particular case of the existence of a compact linear operator which is not nuclear. Funct. Anal. i Pril. 6, 102.

[16] Vachania, N.N.; Tarieladze, V.I.; Čiobanjan, S.A. (1985). Verojatnostnye raspredelenia v Banachovych prostranstvach. Moskva Nauka.

ON SOME PROPERTIES OF A_* , L^1/H^1_0 AS BANACH ALGEBRAS

Juan C. Candeal
Dpto. de An. Económico, E.U.E.E., Zaragoza, Spain

José E. Galé
Dpto. de Matemáticas, Fac. Ciencias, Zaragoza, Spain.

A nuestro profesor Antonio Plans

Abstract. We show some properties of the Banach algebras A_* and L^1/H^1_0 in relation with analytic semigroups and Banach module homomorphisms. In particular we give some results involving the Radon-Nikodym and the analytic Radon-Nikodym properties.

INTRODUCTION.

If $1\le p<+\infty$ we denote by L^p (L^∞, if $p=\infty$) the usual convolution Banach algebra of all p-integrable (essentially bounded) functions on the unit circle $\mathbb{T}$, under the norm $\|\cdot\|_p$ ($\|\cdot\|_\infty$). Write H^p for the closed subalgebra of L^p formed by all functions $f\in L^p$ such that $\hat{f}(n)=0$, $n<0$, where $\hat{f}(n)$ is the n-th Fourier coefficient of f, and put $H^p_0 = \{f\in H^p: \hat{f}(0)=0\}$. Let A_* be the subalgebra of L^1 formed by all functions which are analytic on $\mathbb{D}=\{z\in \mathbb{C}: |z|<1\}$ and continuous on $\bar{\mathbb{D}} = \mathbb{D}\cup\mathbb{T}$.

Endowed with the sup-norm $\|\cdot\|_\infty$ on $\mathbb{T}$, A_* is a Banach algebra. Actually, L^p, H^p, H^p_0, A_* are ideals of L^1. Observe that the injections $f\in A^* \longrightarrow f \in H^p$ and $f\in H^p \longrightarrow \tilde{f}+H^1_0 \in L^1/H^1_0$, where $\tilde{f}(e^{it})=f(e^{-it})$, $t\in[0,2\pi)$, are continuous. The linear structure of the underlying Banach spaces of A_*, L^1/H^1_0 has been extensively studied (see [19], [2], [3], [11],..., for instance). As is well known, the dual Banach spaces of A_*, L^1/H^1_0 are respectively M/H^1_0, H^∞ where M = {complex Borel regular measures on $\mathbb{T}$} is endowed with the norm of the total variation ([19]).

In this note we show some properties of A_*, L^1/H^1_0 considered as convolution Banach algebras. This product in A_* is usually called

Hadamard's product and it satisfies that $\frac{1}{n!}(f*g)^{(n)}(0) = \frac{1}{n!}f^{(n)}(0)\cdot\frac{1}{n!}g^{(n)}(0)$, $f,g\in A_*$, $n\geq 0$. It is not hard to prove that the maximal ideal space of A_* coincides with $\mathbb{N}_0 = \mathbb{N}\cup\{0\}$ and the Gelfand representation is given by

$f\in A_* \longrightarrow \left(\frac{f^{(n)}(0)}{n!}\right)_{n\geq 0} = (\hat{f}(n))_{n\geq 0} \in c_0$. Moreover A_* is a regular non-unital, semisimple Banach algebra satisfying the spectral synthesis ([20], p. 95). The same properties are clearly satisfied by L^1/H_0^1, as a quotient algebra of L^1.

In the first section of this paper we observe that $(a^z)_{Re z>0}$, $a^z(t)=(1-e^{-z}e^{-it})^{-1}$, $t\in[0,2\pi)$, is an analytic semigroup (definitions below) which generates A_* and L^1/H_0^1, and has good growth conditions. We use this semigroup in A_* to give a partial answer in the negative sense to a question of [23]. Also, we give a proof of the fact that the Banach algebras which contain bounded analytic semigroups on $U = \{\mathrm{Re}\ z>0\}$ do not possess the Radon-Nikodym property (RNP). (For definitions and properties of RNP, see [7]). We will also consider a weakened form of RNP, namely the analytic Radon-Nikodym property (ARNP). See definitions and properties in [4], [8].

In the second section we study module homomorphisms from A_* and L^1/H_0^1. If A is a Banach algebra and X is a left Banach A-module, Hom(A;X) stands for the Banach space of all module homorphism from A to X. In [22], Reiffel asked if Hom(A;X) = X whenever A has a bounded approximate identity, X is neounital (i.e. AX = X) and X is a dual Banach space. The first named author has recently proved that the above equality is not true, in general ([5]). Nevertheless, Hom(A;X) = X is obtained in [22], for $A = L^1(G)$, G locally compact group, and X separable. The proof of this is in fact valid for every Banach $L^1(G)$-module X with RNP. We show here that $\mathrm{Hom}(L^1/H_0^1;X) = X$ for every Banach L^1/H_0^1-module X with ARNP. Moreover we prove that 1) $\mathrm{Hom}(L^1/H_0^1;X) = X$, 2) X has ARNP, 3) X is a dual Banach L^1/H_0^1-module, are three equivalent properties if X is separable in its own norm and is a subalgebra of ℓ^∞ (for the coordinatewise product). So it is very clear from this point of view why L^1/H_0^1 fails to have ARNP. A similar result is given for $L^1(G)$, G compact group.

Finally we observe that A_* is an example of a Banach algebra with a quasi-bounded approximate identity (see [26]) whose completion in the multiplier norm is L^1/H_0^1. As a consequence we obtain that $Mul(A_*)$ $(=Hom(A_*,A_*)) = M/H_0^1$. (Actually our interest in characterizing $Mul(A_*)$ motivated this note. But such a characterization was already obtained by A.E. Taylor in 1950! It can be also deduced from [17]. These proofs are different from the one given here).

1. AN ANALYTIC SEMIGROUP GENERATING A_* AND L^1/H_0^1.

Let A be a Banach algebra annd let S be an additive semigroup of complex numbers. A semigroup in A, on S, is a mapping, $z\in S \longrightarrow a^z \in A$ such that $a^{z+w} = a^z \cdot a^w$ for all $z,w \in S$. We say that the semigroup is analytic if S is open and the mapping is analytic. If $S=\mathbb{R}^+$ or S=U the semigroup will be respectively denoted by $(a^s)_{s>0}$, $(a^z)_{Rez>0}$.

Proposition 1.1. *If* $a^z(t)=(1-e^{-z}e^{it})^{-1}$, $t\in [0,2\pi)$, $z\in U$, *then* $(a^z)_{Re\ z>0}$ *is an analytic semigroup which geherates* A_*, H^p, L^1/H_0^1.

Proof. If $z\in U$ and $n\geq 0$, then $\hat{a}^z(n)=e^{-nz}$ hence $a^z * a^w = a^{z+w}$ for all $z\in U$. Moreover, the mapping $z \in U \longrightarrow a^z\in A_*$ is continuous since

$$\sup_{t\in[0,2\pi)} |a^w(t)-a^z(t)| \leq |e^{-w}-e^{-z}| \ (1-e^{-Re\ z})^{-1} \ (1-e^{-Re\ w})^{-1}, \text{ if } z,w \in U,$$

and so the analyticity follows from the vector Morera's theorem.

Let $\langle a^z | Re\ z> 0\rangle^-$ the closed subalgebra of A_* generated by the semigroup. Assume $\mu\in M$ such that $\mu + H_0^1$, considered as a continuous linear functional on A_*, vanishes on $\langle a^z | Re\ z> 0\rangle^-$. Then,

$$0 = \frac{1}{2\pi}\int_0^{2\pi} (1-e^{-z}e^{it})^{-1}d\mu(t) = \frac{1}{2\pi}\int_0^{2\pi} \Big[\sum_{n=0}^{\infty} e^{-nz}e^{int}\Big] d\mu(t) =$$

$$= \sum_{n=0}^{\infty} e^{-nz}\hat{\mu}(-n),$$ for all $z\in U$. Therefore $\hat{\mu}(-n)=0$, if $n\geq 0$, and, by the Riesz Brothers' theorem, we obtain that $\mu\in H_0^1$.

The remainder follows from the continuity of the injections of A_* into H^p, L^1/H_0^1 (they have dense range). □

A. Sinclair suggests in [23] p. 80 this question: if A is a Banach algebra containing an analytic semigroup $(c^z)_{\mathrm{Re}\ z>0}$ such that

$$(*) \qquad \int_{-\infty}^{+\infty} \frac{\log^+\|c^{1+iy}\|}{1+y^2}\, dy < +\infty ,$$

does there exist another semigroup $(b^s)_{s>0}$ in A, bounded on $\mathbb{R}^+$?

The semigroup $(a^z)_{\mathrm{Re}\ z>0}$ of Theorem 1.1 satisfies condition (*), and we will now prove that A_* is a counterexample to the question, in a partial sense. A basic tool for our proof is the following Esterle's result.

Lemma 1.2. ([9], Th. 4.1). *Let* $(b^s)_{s>0}$ *be a semigroup in a separable Banach algebra. Then there exists* $(s_n)_{n\geq 1} \subset \mathbb{R}^+$ *such that* $\lim_n s_n = 0$ *and* $\lim_n \|b^{s+s_n} - b^s\| = 0$ *for every* $s>0$.

Proposition 1.3 . *Let* $(b^s)_{s>0}$ *be a semigroup in* A_* *such that* $E = \{n\in\mathbb{N}_0 :\ \widehat{(b^s)}\neq 0$ *for all* $s>0\}$ *is infinite. Then* $(b^s)_{s>0}$ *is not bounded on* $\mathbb{R}^+$.

Proof. Consider the closed ideal $I = \left[\bigcup_{s>0} b^s A_*\right]^-$ generated by the semigroup $(b^s)_{s>0}$ in A_*. By Lemma 1.2, there is $(s_k)_{k\geq 1} \subset \mathbb{R}^+$ such that $s_k \xrightarrow[k]{} 0^+$ and $b^{s_k} * g \xrightarrow[k]{} g$ for every $g\in I$. So, if $(b^s)_{s>0}$ is bounded on $\mathbb{R}^+$ and $e_k = b^{s_k}$, $k\geq 1$, then $(e_k)_{k\geq 1}$ is a bounded approximate identity in I, hence $I*I=I$ ([1], p. 61). Put $Z(I)=\{n\in \mathbb{N}_0 :\ \hat{g}(n) = 0$ for all $g\in I\}$. As $(b^s)_{s>0}$ is a semigroup, $Z(I)=\mathbb{N}_0\setminus E$. We distinguish two cases

1) There is $f\in I$ such that $\sum_{n\in E} |\hat{f}(n)| = +\infty$. Since $I*I=I$, $f=g*h$ for some $g,h \in I \subseteq A_* \subseteq L^2$; therefore $\hat{f}(n) = \hat{g}(n)\hat{h}(n)$, $n\in E$, with $(\hat{g}(n))_{n\in E}$, $(\hat{h}(n))_{n\in E} \in \ell^2(E)$ whence $\sum_{n\in E} |\hat{f}(n)| < +\infty$, which is a contradiction.

2) For each $f\in I$, $\sum_{n\in E}|\hat{f}(n)|< +\infty$. Since A_* satisfies the spectral synthesis, $I = \ell^1(E) = c_0(E)'$. It follows that $e_k \xrightarrow[k]{} e$ c_0-weakly for some $e\in I$. In particular $\hat{e}_k(n)\xrightarrow[k]{}\hat{e}(n)$, $n\in E$, but $\hat{e}_k(n)\hat{g}(n)\xrightarrow[k]{}\hat{g}(n)$ for all $g\in I$ hence $\hat{e}_k(n)\xrightarrow[k]{} 1$ and so $\hat{e}(n)=1$ for all $n \in E$. It follows that $e \notin I$, a contradiction.□

By means of a duality argument as in part 2) of the above proof, it can be shown that the Banach algebras ℓ^p, H^p $(1\le p<+\infty)$ are also partial counterexamples to the Sinclair's question.

On the other hand, the semigroup $(a^z)_{\mathrm{Re}\ z>0}$ of Proposition 1.1 is bounded in L^1/H^1_0 on U. Thus L^1/H^1_0 serves to illustrate the following general result.

Proposition 1.4. *Let* A *be a non unital Banach algebra containing an analytic semigroup* $(a^z)_{\mathrm{Re}\ z>0}$ *such that* $\sup_U \|a^z\| < +\infty$. *Then* A *does not have* RNP.

Proof. Whithout loss of generality we can assume that A is generated by $(a^z)_{\mathrm{Re}\ z>0}$. Note that $a^z\cdot a^w=a^{z+w}\xrightarrow[z\to iy]{} a^{w+iy}$ if $y\in R$, $w\in U$; therefore there exists $a^{iy}b:= \lim_{z\to iy} a^z b$ for every $b\in A$. It is routine to verify that $a^{-iy}(a^{iy}b)=b$.

Put $C=\{(1+z)^{-1}a^z\colon z\in U\}\cup\{0\}$. The subset C is clearly bounded in A and it is also closed: suppose $(1+z_n)^{-1}a^{z_n}\xrightarrow[n]{} b\in A$. If $|z_n|\xrightarrow[n]{}+\infty$ then $b = 0 \in C$; otherwise, we can assume that $z_n \xrightarrow[n]{} z_0$ for some z_0 such that $\mathrm{Re}\ z_0\ge 0$. If $\mathrm{Re}\ z_0 > 0$ then $b=(1+z_0)^{-1}a^{z_0} \in C$. If $z_0=iy_0$, $y_0\in R$, as $(1+z_n)^{-1}a^{z_n}b'\xrightarrow[n]{} bb'$ for all $b'\in A$, we obtain that $(1+z_0)^{-1}a^{iy_0}b'=bb'$ and so $b'=a^{-iy_0}(a^{iy_0}b')=a^{-iy_0}[(1+z_0)b]b'$ for all $b'\in A$. This is a contradiction because A is non unital.

Now, let F be a real continuous linear functional on A such that $\sup F(C) = \max F(C)$. If $\max F(C) = F(0) = 0$ then $F(C)\le 0$; if $\max F(C) = F((1+z_0)^{-1}a^{z_0})$ for some $z_0\in U$ then the harmonic function $F((1+z)^{-1}a^z)$ attains its maximum on U and so it is constant. In fact, F is null on C since $F((1+z)^{-1}a^z) \xrightarrow[|z|\to +\infty]{} F(0)=0$. In any case, $F(C)\le 0$. Take a character φ of A (A is not radical because of [23], p. 80). Then $\sup_C \operatorname{Re} \varphi = 1$ since $\varphi((1+z^{-1})a^z)\varphi(b) = \varphi((1+z)^{-1}a^z b) \xrightarrow[|z|\to 0]{} \varphi(b)$ for all $b\in A$. Hence Re φ cannot be approximated by real functionals F such that sup F(C) = max F(C). This means that A fails to have RNP ([7], p. 207). □

Proposition 1.4 is an example of the interplay between analytic semigroups and geometric properties. Actually, much more is true: if $\sup_{z\in U, |z|<1} \|a^z\| < +\infty$, then A does not have ARNP ([6]).

2. MODULE HOMOMORPHIMS AND THE BANACH ALGEBRAS A_*, L^1/H_0^1.

Let A be a Banach algebra and let X be a left Banach A-module with respect to the action $(a,x)\in A\times X \longrightarrow a\cdot x\in X$. Recall that if X' is the dual Banach space of X then X' is a left Banach A-module under the action given by $\langle x,a\cdot x'\rangle=\langle a\cdot x,x'\rangle$, $x\in X$, $x'\in X'$, $a\in A$, where $\langle,\rangle$ stands for duality. Then X' is called a <u>left dual Banach A-module</u>.

If $T: A\longrightarrow X$ is a continuous linear operator, we say that T is a <u>right module homomorphism</u> whenever $T(ab)=a\cdot Tb$ for all $a,b\in A$. We denote by Hom (A;X) the Banach space of all right module homomorphisms from A to X endowed with the norm $\|T\|_{op} = \sup_{\|a\|\le 1} \|Ta\|$. In the case X=A we write Hom(A;X) = Mul(A), the right multiplier Banach algebra of A. The spaces Hom(A;X) were introduced in [13]. Representations of Hom (A;X) by means of tensor products have been given in [22], [14]. In the setting of the geometry of Banach spaces, module homomorphisms which are absolutely summing operators have been studied in [15], in several cases. Also, note that the classical Paley's theorem says that $\mathrm{Hom}(A_*;\ell^2) = \ell^2$ ([19], p. 63).

The study of module homomorphisms is connected with the duality in the natural way (see [22], [18], for instante). The proof of the next proposition is well known.

<u>Proposition 2.1.</u> ([22]). *If* A *is a Banach algebra with a bounded approximate identity and* X *is a neounital Banach A-module then* $Hom(A;X')=X'$.

This equality is an isometry and means that $T \in Hom(A;X')$ if and only if there is a unique $x' \in X'$ such that $T(a)=a\cdot x'$ for all $a \in A$.

<u>Remark</u>. By using the Cohen's Factorization Theorem ([1]) we can easily obtain that if A, X are as in Proposition 2.1 and Y is a left Banach A-submodule of X' such that $A\cdot X'=Y$ then $Hom(A;Y) = X'$. This is a way to calculate quickly module homomorphims in some concrete cases: (i) $Mul(L^1(G)) = Hom(L^1(G);M(G)) = M(G)$ where G is a locally compact group, M(G)={complex regular Borel measures on G} and the module action is the convolution, since $L^1(G)*M(G) = L^1(G)$ (this is a classical result in multipliers theory ([16]), (ii) $Mul(c_0) = \ell^\infty$ (the module action is the coordinatewise product). (iii) $Mul(L^1/H^1_0) = Hom(L^1/H^1_0;M/H^1_0) = M/H^1_0$, with the convolution of cosets as module action. (iv) Hom $(L^1/H^1_0,c_0) = Hom(L^1/H^1_0,\ell^\infty) = \ell^\infty$, where $(f+H^1_0)\cdot(x_n)_{n\geq 0} = (\hat{f}(-n)\cdot x_n)_{n\geq 0}$, $f \in L^1$, $(x_n)_{n\geq 0} \in c_0$ (v) $Hom(L^1/H^1_0,A_*) = Hom(L^1/H^1_0,H^\infty) = H^\infty$ etc...

In [22], Rieffel asked if $Hom(A;X)=X$ for a Banach algebra A with a bounded approximate identity and a neounital Banach A-module X which is a dual Banach space. The general answer to this question is no ([5]) but Rieffel showed in [22], p. 477, that $Hom(L^1(G);X) = X$ for X separable. Actually, his proof works for every Banach $L^1(G)$-module with RNP. In the same way, we will now prove that $Hom(L^1/H^1_0;X) = X$ if X has ARNP. We will use a characterization of ARNP given by Dowling. L^∞_X stands for the space of all essentially bounded X-valued functions on $\mathbb{T}$.

<u>Lemma 2.2.</u> ([8], Th. 1). *A Banach space* X *has* ARNP *if and only if for each bounded linear operator* $T:L^1/H^1_0 \longrightarrow X$ *the operator* $T\cdot q:L^1 \longrightarrow X$ *is representable i.e.* $(T\cdot q)(f) = \int_{\mathbb{T}} fg$ *for every* $f \in L^1$ *and some* $g \in L^\infty_X$, *where* $q:L^1 \longrightarrow L^1/H^1_0$ *is the natural quotient operator.*

Theorem 2.3. *Let* X *be a neounital Banach* L^1/H_0^1*-module with* ARNP. *Then* $Hom(L^1/H_0^1;X) = X$.

Proof. If we define $f\cdot x = q(f)\cdot x$, $f\in L^1$, $x\in X$, then X is a neounital left Banach L^1-module. Further, the element $(\mu*f)\cdot y$ of X is independent of the decomposition of $x\in X$ as $x=f\cdot y$ ($f\in L^1$, $y\in X$), for every $\mu\in M$; so X becomes a left Banach M-module under the action $\mu\cdot x = (\mu*f)\cdot y$. Then we define the action of $\mathbb{T}$ on X as $e^{it}\cdot x = \delta_t\cdot x$ where $t\in [0,2\pi)$, δ_t is the Dirac measure at e^{it}, and $x\in X$.

On the other hand, if $T\in Hom(L^1/H_0^1;X)$ then $T_1=T\cdot q$ belongs to $Hom(L^1;X)$ and T_1 is representable by Lemma 2.2 since X has ARNP. Now, by the same argument as in [22], p. 477, we can get $x\in X$ such that $T(q(f)) = T_1(f) = f\cdot x$ for all $x\in X$. □

The converse to the Theorem 2.3 does not hold in general, since $Hom(L^1/H_0^1;\ell^\infty) = \ell^\infty$ or $Hom\ (L^1/H_0^1;M/H_0^1) = M/H_0^1$. However we will give a partial converse in Theorem (2.5). For this we need the following result.

If A is a Banach algebra and X is a left Banach A-module we say that A is X-<u>weakly completely continuous</u> if each operator $T_a(x)=a\cdot x$, $x\in X$, $a\in A$, from X to X is weakly compact.

Lemma 2.4. ([5]). *The following properties are equivalent (here we suppose* A *with a bounded approximate identity):*

(i) A *is* X*-weakly completely continuous.*

(ii) $A\cdot X'' \subset X$.

(iii) $Hom\ (A;X) = X''/R = (A\cdot X')'$, *where* $R = \{x''\in X'':\ a\cdot x''=0$ *for all* $a\in A\}$.

Related results can be seen in [25], [18]. We do not know whether or not the above lemma has also been proved in [27] (see [18], p. 499) because this reference has not been available for us so far.

Property (iii) of Lemma 2.4 is good in order to find some module homomorphisms. Thus $Mul(c_0) = (c_0\cdot\ell^1)' = (\ell^1)' = \ell^\infty$; $Mul(L^1(G)) = (L^1(G)*L^\infty(G))' = C(G)' = M(G)$ and $Hom(L^1(G);C(G)) = (L^1(G)*M(G))' = L^1(G)' = L^\infty(G)$, if G is a compact group; $Mul(L^1/H_0^1) = (L^1/H_0^1*H^\infty)' = (A_*)'$

$= M/H^1_0$; $\mathrm{Hom}(L^1/H^1_0; A_*) = (L^1/H^1_0 * M/H^1_0)' = (L^1/H^1_0)' = H^\infty$, etc.
(see examples in the remark after Proposition 2.1).

Theorem 2.5. *Let X be a separable neounital Banach L^1/H^1_0-module such that L^1/H^1_0 is X-weakly completely continuous. Then the following statements are equivalent:*

(i) X *has* ARNP.

(ii) $\mathrm{Hom}(L^1/H^1_0; X) = X$.

(iii) X *is a dual Banach* L^1/H^1_0*-module.*

In particular this is the case if X *is algebraically isomorphic to a subalgebra of* ℓ^∞ *and the module action is given by* $q(f)*x = (\hat{f}(-n)x_n)_{n\geq 0} \in X$, $f \in L^1$, $x = (x_n)_{n\geq 0} \in \ell^\infty$.

Proof. The first part is clear because of Theorem (2.3) and Lemma (2.4). For the second part, note that the operator $T_p : x \in X \longrightarrow p*x \in X$ is compact for every $p \in L^1/H^1_0$ such that $\hat{p}(n)=0$ whenever $n \geq n_0$, for some n_0. Moreover each operator $T_{q(f)}$, $f \in L^1$, is aproximable by operators T_p in the operator norm topology. Hence L^1/H^1_0 is X-weakly completely continuous. □

As is well known L^1/H^1_0 does not have ARNP. The first proof of this was given in [4], via vector-valued bounded analytic functions. A second proof can be obtained from Lemma (2.2) by taking the operator T=q, $q: L^1 \longrightarrow L^1/H_0$. Our Theorem 2.5 "measures" exactly the lack of ARNP in L^1/H_0, in the setting of the module homomorphims. The same can be said about c_0.

Using similar arguments as above, analogous results can be obtained for the group algebra of a compact group. See [21] for the definition and properties of Segal algebras.

Theorem 2.6. *Let* G *be a compact group and let* X *be a separable neounital left Banach* $L^1(G)$*-module for the convolution action. The following statements are equivalent:*

(i) X *has* RNP.

(ii) $\mathrm{Hom}(L^1(G); X) = X$.

(iii) X *is a dual Banach* $L^1(G)$*-module.*

In particular, this is the case for every Segal algebra X *of*

$L^1(G)$.

Proof. The only thing to be explained is that the irreducible representations of $L^1(G)$, which generate weakly compact operators, are dense in $L^1(G)$ ([12]). □

Any Banach space X is a L^1-module under the module action $f\cdot x = \hat{f}(1)x$ ($f\in L^1$, $x\in X$). If $T\in \mathrm{Hom}(L^1;X)$, $f\in L^1$, and $(e_n)_{n\geq 1}$ is a bounded approximate identity in L^1 we have $T(f) = \lim_n T(f*e_n) = = \lim_n f\cdot Te_n = \lim_n \hat{f}(1)(Te_n)$ whence there is $x\in X$ such that $x = \lim_n Te_n$ and $Tf = \hat{f}(1)x = f\cdot x$. Therefore $\mathrm{Hom}(L^1;X) = X$. This shows that we cannot change the convolution module action in Theorem 2.6, in general.

There is a condition on a Banach algebra A which is less restrictive than that of having a bounded approximate identity and which allows us to identify Mul(A) in many situations, by following a general strategy. This condition consists on the existence in A of a quasibounded right approximate identity, i.e. a right approximate identity $(e_j)_{j\in J}$ such that the family $(T_{e_j})_{j\in J}$ is bounded in the norm of Mul(A) ([26]). The general argument to identify Mul(A) is given by the next result.

Proposition 2.7. ([26], Th. 3.2) *Let A be a semisimple Banach algebra with a quasi-bounded right approximate identity $(e_j)_{j\in J}$. Let B be the completion of A in Mul(A) for the norm $\|\cdot\|_{op}$. Then Mul(A) is topologically isomorphic to Mul(B).*

The interest of Proposition (2.7) is that $(e_j)_{j\in J}$ is a bounded approximate identity in B. Examples for which Proposition 2.7 can be applied, are given in [18], §7. We show here two other interesting examples of this general method. The first one is A_*. The second one is the Banach algebra of all nuclear operators on a Banach space with the bounded approximation property.

Examples.

1) *If $e_r(z) = (1-rz)^{-1}$, $z\in \mathbb{D}$, $0<r<1$, then $(e_r)_{0<r<1}$ is a quasi-bounded approximate identity in A_*. Moreover, the completion of A_* in the operator norm topology is L^1/H_0^1. In particular $\mathrm{Mul}(A_*) = M/H_0^1$.*

To see this, take $f \in A_*$, and put $f_r(z) = f(rz)$ if $z \in \mathbb{D}$, $0<r<1$. Then $f_r \xrightarrow[r\to 1]{} f$ in A_* and $(e_r * f)^{\wedge}(n) = r^n \hat{f}(n) = \hat{f}_r(n)$ for all $n \geq 0$ whence $e_r * f = f_r$. Moreover, $\|e_r\|_{op} = \sup_{\|f\|_\infty \leq 1} \|e_r * f\|_\infty = \sup_{\|f\|_\infty \leq 1} \|f_r\| = 1$ for $0<r<1$.

Now, if $g \in A_*$ then $\|f*g\|_\infty = \|(\tilde{f}+H_0^1)*g\|_\infty \leq \|(\tilde{f}+H_0^1\|_1 \|g\|_\infty$ where $\|\tilde{f}+H_0^1\|_1$ is the norm of $\tilde{f}+H_0^1$ in L^1/H_0^1. Hence $\|f\|_{op} \leq \|f+H_0^1\|_1$. Conversely, $\|f\|_{op} =$

$= \sup_{\|g\|_\infty \leq 1} \|f*g\|_\infty = \sup_{\substack{\|g\|_\infty \leq 1 \\ t \in [0,2\pi)}} |(f*g)(e^{it})| \geq \sup_{\|g\|_\infty \leq 1} |(f*g)(1)| =$

$= \sup_{\|g\|_\infty \leq 1} \left| \frac{1}{2\pi} \int_0^{2\pi} \tilde{f}(e^{is}) g(e^{is}) ds \right| = \|\tilde{f}+H_0^1\|_1$ (The last equality holds because of the duality between A_* and M/H_0^1).

The isomorphism $Mul(A_*) = M/H_0^1$ was proved by the first time by A.E. Taylor in 1950 ([24], Th. 9.3). It is also a consequence of the main theorem of [17], although it was not mentioned there.

2) Let X be a Banach space and put $(x' \otimes x)(y) = \langle y, x' \rangle x$ for all $x, y \in X$, $x' \in X'$. Let $\mathcal{N}(X)$ be the Banach algebra of the *nuclear operators* on X, i.e. $\mathcal{N}(X) = \{N: X \longrightarrow X \mid N = \sum_{n=1}^{\infty} x'_n \otimes x_n,\ x_n \in X,\ x_n' \in X'$ and $\sum_{n=1}^{\infty} \|x'_n\| \ \|x_n\| < +\infty\}$ endowed with the norm $\|N\|_1 = \inf \left\{ \sum_{n=1}^{\infty} \|x'_n\| \ \|x_n\| : N = \sum_{n=1}^{\infty} x'_n \otimes x_n \right\}$. It is suitable for us to write $ST = T \cdot S$ where $S, T \in \mathcal{L}(X) =$ {bounded linear operators on X} and "$\cdot$" stands for composition. Then each element of $\mathcal{L}(X)$ defines by this reversed composition a right multiplier of $\mathcal{N}(X)$. If we write $\||T|\| = \sup_{\|N\|_1 \leq 1} \|NT\|_1$ then $\||T|\| \leq \|T\|_{op}$ since $\|NT\|_1 \leq \|T\| \ \|N\|_1$. Conversely, if $x \in X$, $x' \in X'$, and $\|x\|$, $\|x'\| \leq 1$ we have $\|Tx\| = \|x' \otimes Tx\|_1 = \|T_0(x' \otimes x)\|_1 \leq \||T|\|$, whence $\||T|\| = \|T\|$. So, the $\||\cdot|\|$-completion of $\mathcal{N}(X)$ equals its $\|\cdot\|_{op}$-completion, i.e. the $\|\cdot\|_{op}$-completion of finite rank operators, say F(X).

Now $Mul(F(X)) = \mathcal{L}(X)$ ([13]) and F(X) has a right bounded approximate identity since X has the bounded approximation property (in this case, F(X) equals the Banach algebra K(X) of the compact operators on X ([10], p. 119)). Thus, with respect to Proposition (2.7),

"*If* X *is a Banach space with the bounded approximation property then* A = $\mathcal{N}(X)$ *has a quasibounded approximate identity*, B = K(X) *and* Mul($\mathcal{N}$(X)) = $\mathcal{L}$(X)".

Observe that this result cannot be attained by duality techniques. It generalizes the cases X reflexive ([18]) and X = H, H a Hilbert space ([17]) for which $\mathcal{N}$(H) = {trace-class operators on H}.

Finally, we note that the above ideas when applied to the Banach module case imply more generally that Hom(A; X) equals Hom(B; Y) where A, B are as above and Y is the Hom(A; X)-completion of X. The Paley's theorem ([19]) may be viewed as an example of this:
$\mathrm{Hom}(A;\ell^1) = \mathrm{Hom}(L^1/H_0;\ell^2) = \ell^2$ (the question here is to find that the norm in $\mathrm{Hom}(A_*;\ell^1)$ is the ℓ_2-norm). If we associate the Paley's theorem to the results of [22] and [14] we get that the following properties are equivalent:

(i) $\mathrm{Hom}(A_*;\ell^1) = \ell^2$; (ii) $\ell^2 = A_* \hat{\otimes} c_0/k_1$, where $k_1 = \overline{\mathrm{span}}\,\{fg \otimes x - g \otimes f x \mid f,g \in A_*,\ x \in c_0\}$; (iii) $\ell^2 = A_* \hat{\otimes} c_0/k_2$ where $k_2 = \ker P$ and $P\left(\sum_n f_n \otimes x_n\right) = \sum_n f_n \cdot x_n$, $f_n \in A_*$, $x_n \in c_0$ (here $A_* \hat{\otimes} c_0$ is the projective tensor product of A_* and c_0): (i) $\Leftrightarrow$ (ii) because $\mathrm{Hom}(A_*;\ell^1) = (A_* \hat{\otimes} c_0/k_1)'$ ([22], Th. 3.21); (ii) $\Leftrightarrow$ (iii) because A_* has a quasibounded approximate identity ([14], p. 5). In particular, the mapping $f \otimes x \in A_* \hat{\otimes} c_0 \longrightarrow f \cdot x \in \ell^2$ is surjective.

REFERENCES.

[1] Bonsall, F.F. & Duncan, J. (1973). Complete Normed algebras, Springer-Verlag.

[2] Bourgain J. (1984). New Banach space properties of the disc algebra and H^∞, Acta Math. 152, 1-48.

[3] Bourgain, J. (1984). Bilinear forms on H^∞ and bounded bianalytic functions, Trans. Amer. Math. Soc. 286 (1), 313-337.

[4] Bukhalov, A.V. & Danilevich, A.A. (1982). Boundary properties of analytic and harmonic functions with values in a Banach space, Math. Notes 31, 104-110.

[5] Candeal, J.C. (1988). Doctoral Dissertation, Zaragoza.

[6] Candeal, J.C. & Galé, J.E. On the existence of analytic semigroups bounded on the half-disc in some Banach algebras. (Preprint).

[7] Diestel, J. & Uhl, J.J. (1977). Vector Measures. Mathematical Surveys 15, AMS.

[8] Dowling, P.N. (1985). Representable operators and the analytic Radon-Nikodym property in Banach spaces, Proc. Royal Irish Acad. 85 143-150.

[9] Esterle, J. (1983). Elements for a classification of commutative radical Banach algebras, Lecture Notes in Math. 975, Springer-Verlag.

[10] Esterle, J. (1984). Mittag-Leffler methods in the theory of Banach algebras and a new approach to Michael's problem, Contemporary Math. AMS vol. 32, 107-130.

[11] Fernandez, J.L. A boundedness theorem for L^1/H^1_0, Preprint.

[12] Hewitt, E. & Ross, K.A. (1970). Abstract Harmonic Analysis II, Springer-Verlag.

[13] Johnson, B.E. (1964). An introduction to the theory of centralizers, Proc. London Math. Soc. (3) 14, 299-320.

[14] Johnson, D.L. & Lahr, Ch. D. (1982). Weak approximate identities and multipliers, Studia Math. LXXIV, 1-15.

[15] Kwapien, S. & Pelczynski, A. (1978). Remarks on absolutely summing translation invariant operators from the disc algebra and its dual into a Hilbert space, Michigan Math. J. 25 , 173-181.

[16] Larsen, R. (1971). An Introduction to the Theory of Multipliers, Springer-Verlag.

[17] Oshoby, E.O. & Pym, J.S. (1981). Banach algebras whose duals are multiplier algebras, Bull. London Math. Soc. 13, 66-68.

[18] Oshoby, E.O. & Pym, J.S. (1987). Banach algebras whose duals consist of multipliers, Math. Proc. Camb. Phil. Soc. 102, 481-505.

[19] Pelczynski, A. Banach Spaces of Analytic Functions and Absolutely Summing operators, Reg. Conf. Series in Math. 30, AMS, Providende, Rhode Island.

[20] Porcelli, P. (1966). Linear Spaces of Analytic Functions, Rand. McNally.

[21] Reiter, H. (1988). Classical Harmonic Analysis and Locally Compact Groups, Oxford Univ. Press.

[22] Riefel, M.A. (1967). Induced Banach representations of Banach algebras and locally compact groups, J. Funct. Anal. 1, 443-491.

[23] Sinclair, A. (1982). Continuous Semigroups in Banach Algebras, London Math. Soc., Lecture Notes Series 63, Cambridge Univ. Press.

[24] Taylor, A.E. (1950). Banach spaces of analytic functions in the unit circle I, Studia Math. XI, 146-170.

[25] Tomiuk, B.J. (1981). Arens regularity and the algebra of double multipliers, Proc. AMS 81, 293-298.

[26] Tomiuk, B.J. (1986). Isomorphisms of multiplier algebras, Glasgow Math. J. 28, 73-77.

[27] Grosser, M. (1977). Bidualräume und Vervollstanddigungen von Banachmoduln, Lecture Notes in Math. 717 Springer-Verlag.

Acknowledgements. We would like to thank Professors G.R. Allan and J. Esterle for talks about subjects of this paper.

This research has been supported by the Spanish CAYCIT Grant number 0804-84 and the Secretaría de Relaciones Internacionales de la Universidad de Zaragoza, Spain.

ON FACTORIZATION OF OPERATORS

M. González
Univ. of Santander, Spain

V.M. Onieva
Univ. of Zaragoza, Spain

Abstract. We present a construction method of operator ideals based on certain functions S from the class B of all Banach spaces into the class N of all normed spaces; examples of such functions have been considered in the literature. With additional assumptions on S and by applying the factorization of Davis-Figiel-Johnson-Pelczynski and another factorization technique dual in a sense of the above, we are able to prove the factorization property for such ideals; the well-known result about weakly compact operators is covered.

The well-known factorization of Davis-Figiel-Johnson-Pelczynski [2], briefly DFJP factorization, shows how an operator in the class $\mathcal{L}$ of all bounded linear operators between Banach spaces, say $T \in \mathcal{L}(E,X)$, factors through a Banach space Y in such a way that in the corresponding product $T = jA$ the operator j is tauberian injective; moreover Y is reflexive if and only if T is weakly compact. This construction has been used by many authors and systematically studied in [13], [14].

In [4] the authors presented another factorization $T = Uk$ through a Banach space Z, dual in a sense of the above.

In this paper we give characterizations, in terms of the properties of Y and Z, for some operator ideals defined by means of certain functions of B into N, analogous to the above characterization of weakly compact operators.

We shall use standard notations and canonical isometric identifications; for example, sometimes we shall regard any Banach space E as the subset $J_E E$ of E" being J_E the canonical embedding of E into E"; we also write J(E) instead of $J_E E$. Moreover for $E, X \in$ B and $T \in \mathcal{L}(E,X)$ we denote $H(E) := E''/E$, H(T) the operator in $\mathcal{L}(H(E),H(X))$ induced by the biconjugate T" of T in the way

$$H(T):\ x'' + E \in H(E) \longrightarrow T''x'' + X \in H(X),$$

$\hat{T}$ the injective operator associated with T, and q the quotient map from E onto E/N(T), so that $T = \hat{T}q$.

Recall that $T\in \mathcal{L}(E,X)$ is tauberian provided that $(T'')^{-1}(X)\subset E$; of course $T''(E)\subset X$, so actually T is tauberian if and only if we have $(T'')^{-1}(X)= E$, [9], [17], or equivalently H(T) injective. Then, T is said to be a <u>cotauberian</u> operator provided that H(T) has range dense in H(X). It is clear that T is cotauberian if and only if T' is tauberian.

For informations and notations about operator ideals we refer to the classical reference work of A. Pietsch [15]. But we recall that with every operator ideal U there is associated the Banach space ideal $Sp(U):= \{E\in B \mid I_E\in U\}$ where I_E is the identity map of E, and with every Banach space ideal A there is associated the operator ideal $Op(A):=\{T\in\mathcal{L} \mid T$ factors through some $E \in A\}$; then we have $A = Sp(Op(A))$, but $Op(Sp(U))\subset U$. The factorization property for an operator ideal U means $U = Op(Sp(U))$. On the other hand WCo, WCC and UC will denote the operator ideals of all weakly compact, weakly completely continuous and unconditionally convergent operators, respectively; and R, WSC and Nc_0 will be the Banach space ideal of reflexive, weakly sequentially complete and with no copy of c_0 respectively.

Next, we give a brief description of the tauberian and cotauberian factorizations of an operator.

Let E and X be Banach spaces, $T\in \mathcal{L}(E,X)$. For each n=1,2,... the gauge $\|\cdot\|_n$ of the set $2^nT(B_E)+2^{-n}B_X$ is a norm equivalent to $\|\cdot\|$; denote $X_n:=(X,\|\cdot\|_n)$ and define for $x\in X$, $\||x|\|:=\left(\sum_{n=1}^{\infty}\|x\|_n^2\right)^{1/2}$; let $Y:=\{x\in X,\ \||x|\|<\infty\}$. Then $(Y,\||\cdot|\|)$ is a Banach space, the identity embedding j of Y into X is an injective tauberian operator such that $A:= j^{-1}T\in \mathcal{L}(E,Y)$, and $T=jA$ is the DFJP factorization or tauberian factorization of T.

On the other hand, $p_n(x):=2^n\|Tx\|+2^{-n}\|x\|$ defines a norm in E equivalent to $\|\cdot\|$, let $E_n:=(E,p_n)$. Consider

$$D:=\{(f_n)\subset E' \mid f_n=f_1,\ n=1,2,\dots\},$$

$$i: x \in E \longrightarrow (x,0,0,\dots) \in \ell_2(E_n),$$

$$P: (x_n) \in \ell_2(E_n) \longrightarrow (x_n)+{}^0D \in \ell_2(E_n)/{}^0D.$$

The intermediate space of the cotauberian factorization of T is $Z := \ell_2(E_n)/{}^0D$ and $k := Pi \in \mathcal{L}(E,Z)$ is one of its factors; we have $N(k) = N(T)$, $R(k)$ dense and k cotauberian [4]. The second factor $u \in \mathcal{L}(Z,X)$ is obtained as the continuous extension of $\hat{T}\hat{k}^{-1}$ where $T = \hat{T}q$ and $k = \hat{k}q$ are the canonical decompositions with q the quotient map $q: E \longrightarrow E/N(T)$. We can write $T = Uk$, the cotauberian factorization of T, [4].

The construction of the intermediate space Y of the tauberian factorization of an operator $T \in \mathcal{L}(E,X)$ can be removed of the operational context replacing $T(B_E)$ by a bounded absolutely convex subset W of X as starting points. For the cotauberian factorization it is possible as well starting with a continuous seminorm p in E and defining $p_n(x) := 2^n p(x) + 2^{-n}\|x\|$, $x \in E$.

Now we present the special functions S of B into N which generate two operator ideals U^S and U_S; the study of properties of these ideals is our goal.

1. Definition. Let S be a function from the class B of all Banach spaces into the class N of all normed spaces. We shall say that S is an ideal function if the following conditions are satisfied:

(I-1) $J(E) \subset S(E) \subset E''$ for every $E \in B$.

(I-2) $T''(S(E)) \subset S(X)$ for every $E, X \in B$ and $T \in \mathcal{L}(E,X)$.

Moreover, S is said to be closed if, in addition,

(I-3) $S(E) \in B$ for every $E \in B$.

We shall say that S satisfies the subspace condition if

(I-S) $i''(S(M)) = S(E) \cap M^{00}$ for every $E \in B$ and every subspace M of E (i is the inclusion map of M into E);

and S satisfies the quotient condition if

(I-Q) $q''(S(E)) = S(E/M)$ for every $E\in B$ and every subspace M of E (q is the quotient map $q: E \longrightarrow E/M$).

Clearly $J: E\in B \longrightarrow J(E)\in B$ and $(''): E\in B \longrightarrow E''\in B$ are closed ideal functions with the properties I-S and I-Q.

Other examples of ideal functions are defined by the following subspaces of E'':

DSC(E): w^*-limits in E'' of weakly unconditionally Cauchy series in E.

$B_1(E)$: w^*-limits in E'' of w^*-Cauchy sequences in E.

$GS(E) := \bigcup \{M^{oo} \mid M$ separable subspace of $E\}$.

B_1 is closed [11] and DSC is not closed [1], and both functions satisfy the subspace condition [12]; they do not satisfy the quotient condition: $DSC(\ell_1) = B_1(\ell_1) = J(\ell_1)$ because ℓ_1 is weakly sequentially complete, but $J(c_0) \neq DSC(c_0) \subset B_1(c_0)$. On the other hand GS is closed, and it satisfies the quotient condition: in fact, let M be a subspace of $E\in B$, q the associated quotient map, and F a separable subspace of E/M. Then F is isomorphic to Q/M for some subspace Q of E which contains M; as Q/M is separable, by [16; lemma 2] there exists a separable subspace V of Q such that $V+M = Q$, so $qV = F$, hence $q''V^{oo} = F^{oo}$, and we infer $q''(GS(E)) = GS(E/M)$.

The following simple result provides information about the behaviour of general ideal functions in relation with subspaces M and quotients E/M of $E\in B$.

2. <u>Proposition</u>. *Let* S *be any ideal function,* $E, M\in B$, M *subspace of* E. *Then:*

(i) $i''(S(M)) \subset S(E)\cap M^{oo}$; $q''(S(E)) \subset S(E/M)$.

(ii) *If* M *is complemented,* $i''(S(M)) = S(E)\cap M^{oo}$ *and* $q''(S(E)) = S(E/M)$.

<u>Proof</u>. (i) It follows from property I-2 of S; note that $R(i'') = M^{oo}$.

(ii) As M is complemented in E, there exist $P\in \mathcal{L}(E/M)$ and $Q\in \mathcal{L}(E/M, E)$ such that iP is projector onto M and Qq is projector onto $N(iP)$ with kernel $N(Qq) = M$. Then $i''P''$ and $Q''q''$ are projectors onto M^{oo} and $N(iP)^{oo} \equiv E''/M^{oo} \equiv$

$\equiv(E/M)''$ respectively, and by virtue of the property I-2 of S applied to P, Q we have

$$i''(S(M)) \subset S(E) \cap M^{oo} = i''P''(S(E)) \cap M^{oo}) \subset i''(S(M)),$$
$$q''(S(E)) \subset S(E/M) = q''Q''(S(E/M)) \subset q''(S(E)).$$

Next we shall associate two operator classes with an ideal function.

3. Definition. Given an ideal function S, we define operator classes U^s and U_s by means of their components for $E, X \in B$ in the following way:

$$U^s(E,X) := \{T \in \mathcal{L}(E,X) \mid T''(S(E)) \subset J(X)\},$$
$$U_s(E,X) := \{T \in \mathcal{L}(E,X) \mid T''(E'') \subset S(X)\}.$$

We collect now some elementary properties of U^s and U_s.

4. Proposition. *Let S be an ideal function. The following statements hold:*

(i) U^s *and* U_s *are operator ideals such that* U^s *is injective and closed,* U_s *is surjective and* $U^s \cdot U_s = WCo$.

(ii) *If S closed, then* U_s *is closed and* $U^s \cap U_s = WCo$.

(iii) $Sp(U^s) = \{E \in B \mid S(E) = J(E)\}$, $Sp(U_s) = \{E \in B \mid S(E) = E''\}$ and $Sp(U^s) \cap Sp(U_s) = R$.

Proof. (i) Clearly U^s, U_s are ideals and $U^s \cdot U_s = WCo$.

Now, let $T \in \mathcal{L}(E,F)$ and $i \in \mathcal{L}(F,X)$ an injection such that $iT \in U^s$, that is, $i''T''(S(E)) \subset J(X)$. Since i is tauberian [16;11.4.5], it follows that $J(F) \subset i''^{-1}(J(X))$, so we have $T''(S(E)) \subset J(F)$ and then $T \in U^s$; thus U^s is injective. Furthermore, if $(T_n) \subset U^s(E,F)$ with $\|T_n - T\|$ convergent to 0, we have $T_n''f \longrightarrow T''f \in J(F)$ for every $f \in S(E)$; so $T''(S(E)) \subset J(F)$ and $T \in U^s$.

On the other hand , if $T \in \mathcal{L}(F,X)$ and $q \in \mathcal{L}(E,F)$ is a surjection such that $Tq \in U_s$, then $T \in U_s$ because $R(T''q'') = R(T'') \subset S(X)$; thus U_s is surjective.

(ii) It is clear that S closed implies U_s closed, and in such a case $WCo = U^s \cap U_s$, [9].

(iii) It is obvious.

Our next result is related with the intermediate spaces of the tauberian and cotauberian constructions; but we shall need two lemmas.

5. <u>Lemma</u>. *Let* S *be a closed ideal function. For every sequence* (E_n) *of Banach spaces we have* $S(\ell_2(E_n)) = \ell_2(S(E_n))$.

<u>Proof</u>. As E_m is complemented in $\ell_2(E_n)$ we have

$$(*)\quad i''(S(E_m)) = S(\ell_2(E_n)) \cap E''_m$$

where we identify E''_m with the subspace $\{(f_n \in \ell_2(E''_n) \mid f_n=0 \text{ if } n\neq m\}$ of $\ell_2(E''_n)$. If P_m is the m-th coordinate projectionn of $\ell_2(E_n)$, then P''_m is the m-th coordinate projection of $\ell_2(E''_n)$. By (*) we have

$$(f_n) \in S(\ell_2(E_n)) \Rightarrow P''_m((f_n)) = f_m \in E_m \text{ for every } m \Rightarrow$$
$$\Rightarrow (f_n) \in \ell_2(S(E_n)),$$

and

$$(f_n) \in \ell_2(S(E_n)) \Rightarrow (0,\dots,0,f_m,0,\dots) \in S(\ell_2(E_n)) \text{ for every } m$$
$$\Rightarrow (f_n) \in S(\ell_2(E_n));$$

note that the fact that S is closed has been employed in the last implication.

The following proposition is known in the literature, and gives a description of the injective or surjective closed envelope $\overline{U}^{inj}$ and $\overline{U}^{sur}$ of an operator ideal U.

6. <u>Lemma</u>. *Let* U *be an operator ideal,* $E, X \in B$ *and* $T \in \mathcal{L}(E,X)$.

(a) T *belongs to* $\overline{U}^{inj}$ *if and only if there are a Banach space* H, *an operator* $V \in U(E,H)$, *and a function* g *of* $(0,\infty)$ *into itself such that*

$$\|Tx\| \leq g(\varepsilon)\|Vx\| + \varepsilon\|x\|,\ \forall x \in X,\ \forall \varepsilon>0 \text{ [7;20.7.3], [8;2.9].}$$

(b) T *belongs to* $\overline{U}^{sur}$ *if and only if there are a Banach space* H, *an operator* $V \in U(H,X)$ *and a function* g *of* $(0,\infty)$ *ito itself such that*

$$T(B_E) \subset g(\varepsilon)V(B_H) + \varepsilon B_X,\ \forall \varepsilon>0, \text{ [8;2.9].}$$

Next we are going to present the result mentioned above.

7. <u>Theorem</u>. *Let S be an ideal function, $E, X \in B$, W a bounded absolutely convex subset of X and p a continuous seminorm in E.*

(i) *If S is closed and has the subspace property and W^{oo} is contained in S(Y), then the intermediate space Y of the tauberian construction belongs to $Sp(U_s)$, that is, $S(Y)=Y''$.*

(ii) *If S has the quotient property and $\{x \in E | p(x) \leq 1\}^o$ is relatively $\sigma(E', S(E))$-compact, then the intermediate space Z of the cotauberian construction belongs to $Sp(U^s)$, that is, $S(Z) = J(Z)$.*

<u>Proof</u>. (i) Let $j: Y \longrightarrow X$ be the identity embedding; it is well known that $j = P\varphi$ where $\varphi: y \in Y \longrightarrow (j(y), j(y), \ldots) \in \ell_2(X_n)$ is an isometry and P is the first coordinate projection.

Firstly we shall prove

$$(*) \quad f \in Y'' \text{ and } j''f \in S(X) \text{ implies } f \in S(Y).$$

For this, consider $j''f = P''\varphi''f \in S(X)$ ($=S(X_n)$ for every n); as $S(\ell_2(X_n)) = \ell_2(S(X_n))$, we have $\varphi''f = (j''f, j''f, \ldots,) \in \ell_2(S(X_n))$ and

$$\varphi''f \in R(\varphi'') \cap S(\ell_2(X_n)) = Y^{oo} \cap S(\ell_2(X_n)) = \varphi''(S(Y));$$

it follows that $f \in S(Y)$ since φ'' is isometry.

Take $E \in B$ and $A \in \mathcal{L}(E, X)$ with $A(B_E) = W$. Since $A''(B_{E''}) \subset W^{oo} \subset S(X)$, we have $A \in U_s$. Moreover $j(B_Y) \subset 2^n A(B_E) + 2^{-n} B_X$ for every n. As U_s is surjective and closed, using lemma 6-(b) we conclude that $j \in U_s$, thus $R(j'') \subset S(X)$. Therefore $Y''=S(Y)$, by (*).

(ii) Consider the cotauberian operator of the cotauberian construction with dense range $k \in \mathcal{L}(E, Z)$, and let H be the completion of $(E/N(p), p)$. The operator $T \in \mathcal{L}(E, H)$ defined by means of $T: x \in E \longrightarrow Tx := x + N(p) \in H$, satisfies $\|Tx\| = p(x)$ for every $x \in E$, so $\{x \in E | \; p(x) \leq 1\} = T^{-1}(B_H)$ and then $\{x \in E | \; p(x) \leq 1\}^o = T'(B_{H'})$ is relatively $\sigma(E', S(E))$-compact [9; 32:1.8-10] or equivalently $T''(S(E)) \subset J(E)$ [6; th 2.7], that is, $T \in U^s(E, H)$.

On the other hand we have

$$(**) \quad \|kx\| \leq 2^n \|Tx\| + 2^{-n} \|x\| \text{ for } x \in E \text{ and every n.}$$

In fact, we know that $k = Pi$ where $i: x \in E \longrightarrow (x0, 0\ldots) \in \ell_2(E_n)$ and

$P:(x_n)\in\ell_2(E_n)\longrightarrow (x_n)+{}^0D\in \ell_2(E_n)/{}^0D$, D the diagonal subspace of $\ell_2(E'_n)$. Since $(-x,0,\dots,0,x,0,\dots)\in {}^0D$ and the norm of $(0,\dots,0,x,0,0,\dots)$ is $2^n\|Tx\|+2^{-n}\|x\|$, we deduce (**) by definition of the quotient norm.

Now, as U^s is a closed injective ideal, from $T\in U^s$ and (**) we derive $k\in U^s$ by virtue of lemma 6-(a), that is, $k''(S(E)()\subset J(Z)$. Therefore to complete the proof, it is enough to show that $k''(S(E))$ is dense in $S(Z)$.

To this end, note that $k'' = P''i''$ where i'' is the embedding $i'':f\in E''\longrightarrow(f,0,0,\dots)\in \ell_2(E''_n)$ and P'' is the quotient mapping $P'':(f_n)\in\ell_2(E''_n)\longrightarrow(f_n)+D^0\in \ell_2(E''_n)/D^0$; the quotient property of S guarantees that $P''(S(\ell_2(E_n)) = S(\ell_2(E_n)/D^0)\equiv S(Z)$. Then, given $g\in S(Z)$ there exists $(f_n)\in S(\ell_2(E_n))$ such that $P''((f_n))=(f_n) + D^0 = g$, and if P_m is the m-th coordinate projection of $\ell_2(E_n)$ we have $P''_m((f_n))=f_m\in S(E_m)\equiv S(E)$ by virtue of the property I-2 of S.

We now take $f_1+\dots+f_m\in S(E)$; by observing that

$$g_1:=(-(f_2+\dots+f_m),f_2+\dots+f_m,0,0,\dots)\in D^0$$
$$g_2:=(0,-(f_3+\dots+f_m),f_3+\dots+f_m,0,0,\dots)\in D^0$$
$$\dots\dots\dots\dots\dots\dots\dots\dots\dots\dots\dots\dots$$
$$g_{m-1}:=(0,0,\dots,-f_m,f_m,0,0,\dots)\in D^0,$$

we have

$$\|k(f_1+\dots+f_m)-g\| = \|(f_1+\dots+f_m,0,0,\dots)+g_1+\dots+g_{m-1}+D^0-g\| =$$
$$= \|(f_1,\dots,f_m,0,0,\dots)+D^0-g\| = \|(0,\dots,0,f_{m+1},f_{m+2},\dots)+D^0\|\le$$
$$\le \|(0,\dots,0,f_{m+1},f_{m+2},\dots)\|,$$

and consequently $\lim_m \|k(f_1+\dots+f_m)-g\|=0$. Thus we conclude that $k''(S(E))$ is dense in $S(Z)$.

An important consequence of the above theorem states that with the same assumptions on S the ideals U_s, U^s have the factorization property.

8. Theorem. *Let S be an ideal function.*

(i) *If S is closed and satisfies the subspace condition,* U_s *has the factorization property, that is* $U_s = Op(Sp(U_s))$.

(ii) *If* S *satisfies the quotient condition, then* U^S *has the factorization property, that is* $U^S = Op(Sp(U^S))$.

Proof. (i) First note that for $T\in L(E,X)$ we have $T''(B_{E''})=T(B_E)^{00}$ because $B_E^{00} = B_{E''}$ is the $\sigma(E'',E')$-closure of B_E in E", and analogously $T(B_E)^{00}$ is the $\sigma(X'',X')$-closure of $T(B_E)$ in X"; since T" is weak*-continuous and $B_{E''}$ weak*-compact, the set $T''(B_{E''})$ is weak*-compact. Because $T''(B_{E''})\subset T(B_E)^{00}$ we can deduce that $T''(B_{E''})=T(B_E)^{00}$.

Now, suppose that $T\in U_s$, i.e. $R(T'') = span(T''(B_{E''})) = span(T(B_E)^{00}) \subset S(X)$. By virtue of the above theorem with $W:= T(B_E)$, we conclude that T factors through $Y\in Sp(U_s)$. Thus $U_s\subset Op(Sp(U_s))$ and therefore U_s has the factorization property.

(ii) Suppose that $T\in U^S(E,X)$, i.e. $T''(S(E))\subset J(X)$, and consider the continuous seminorm p in E given by $p(x):=\|Tx\|$, $x\in E$. Then $\{x\in E|\ p(x)\le 1\}^0 = T^{-1}(B_X)^0 = T'(B_{X'})$ is relatively $\sigma(E',S(E))$-compact [6] and by the above theorem T factors through Z in $Sp(U^S)$. Thus $U^S = Op(Sp(U^S))$.

In relation with the above theorem we note that there are conditions on the function S that do not guarantee the factorization property of the operator ideal U^S or U_s. For example, the ideals U^{B_1}=WCC and U^{DSC} = UC [6] do not have the factorization property since in [3] an operator $T_1\in \mathcal{L}(X_1,c_0)$ is given, where X_1 is a Banach lattice, such that $T_1 \in$ WCC and $T_1\notin Op(Nc_0)$, hence $T_1\in$ WCC \ Op(WSC) and $T_1\in$ UC \ $Op(Nc_0)$ since WCC$\subset$ UC and Op(WSC) is contained in $Op(Nc_0)$.

Finally it is obvious that 8i and 8-ii cover the well-known result of factorization of weakly compact operators.

REFERENCES

[1] Bourgain, J. (1980). Remarks on the double dual of a Banach space. Bull. Soc. Math. Belg. 31, 171-178.

[2] Davies, W.J.; Figiel, T.; Johnson, W.B. & Pelczynski, A. (1974). Factoring weakly compact operators. J. Funct. Anal. 17, 311-327.

[3] Ghoussoub, N. & Johnson, W.B (1984). Counterexamples to several problems on the factorization of bounded linear operators. Proc. AMS 92, 233-238.

[4] Gomzalez, M. & Onieva,(1988). V.M. Duality and factorization of operator ideals. (Preprint).

[5] Heinrich, S. (1980). Closed operator ideals and iterpolation. J. Funct. Anal. 35, 397-411.

[6] Howard, J. & Melendez, K. (1977). Characterizating operators by their first and second adjoints. Bull. Inst. Math. Acad. Sinica 5 , 129-134.

[7] Jarchow, H. (1981). Locally convex spaces. B.G. Teubner.

[8] Jarchow, H.. (1986).Weakly compact operators on C(K) and C^*-algebras. Lectures given at L'Ecole d'Automne CIMPA Nice.

[9] Kalton, N. & Wilansky, A. (1976). Tauberian operators in Banach spaces. Proc. AMS 57, 251-255.

[10] Kothe, G. Topological vector spaces II. Springer Verlag 1979.

[11] McWilliams, R.D. (1968). On the w^*-sequential closure of a Banach space in its second dual. Duke Math. J. 35, 369-373.

[12] Melendez,(1973). K. Sequential properties in Banach spaces. Doctoral Diss., Oklahoma St. Univ.

[13] Neidinger, (1984).R.D. Properties of tauberian operators on Banach spaces. Dotoral Diss., Univ. of Texas at Austin.

[14] Neidinger, R.D. (1984). Factoring operators through hereditarily-ℓ_p spaces. Springer Verlag, LNM 1166, 116-128.

[15] Pietsch,(1980) A. Operator ideals. North-Holland.

[16] Valdivia, M.(1977). On a class of Banach spaces. Studia Math. 60, 11-13.

[17] Wilansky, A. (1978). Modern methods in topological vector spaces. McGraw Hill.

A.M.S. Classification (1980): 46B20, 47D30.

SOME PROPERTIES OF BANACH SPACES Z^{**}/Z.

M. Valdivia
Fac. de Matemáticas, Burjasot (Valencia), Spain

A mi buen amigo Antonio Plans

Abstract. Let (X_n) be a sequence of separable and reflexive infinite dimensional Banach spaces. It is proved in this paper that for every separable Banach space X there is a Banach space Z with the following properties: Z^{**}/Z is norm-isomorphic to X, Z has a monotone and shrinking Schauder decomposition (Z_n) such that Z_n is isomorphic to X_n, $n=1,2,\dots$; Z^* has a monotone and shrinking Schauder decomposition (Y_n) with Y_n isomorphic to X_n^*, $n=1,2,\dots$; Z^{***}/Z^* is norm-isomorphic to X^*. Another results on weakly compactly generated Banach spaces of the kind Y^{**}/Y are also given.

1. NOTATIONS AND BASIC CONSTRUCTIONS.

The vector spaces we shall use here are defined on the field K of the real or complex numbers. Let V be any vector space. If $x=(x_n)$ is a sequence in V and r is a positive integer, we denote by x[r] the sequence (y_n) of V such that $x_n=y_n$, $n=1,2,\dots,r$, $y_n=0$, $n=r+1,r+2,\dots$, and by x{r} the sequence (z_n) of V with $z_n=0$, $n\neq r$ and $z_r=x_r$. If A is a subset of V, L(A) is the linear hull of A. We denote by I the identity operator on V.

Let X be a Banach space. We shall write X^* for its topological dual (accordingly, X^{**} and X^{***} will denote the topological dual of X^* and X^{**}, respectively). As it is usual, we shall consider X as subspace of X^{**}. For every subset Y of X, $\tilde{Y}$ denotes its weak-star closure in X^{**}. If x belongs to X and u to X^*, then $\langle x,u\rangle$ means u(x). If A is an absolutely convex and bounded subset of X, we set X_A for the normed space on L(A) with norm the gauge of A; A^o is the polar of A in X^*. We shall use the same notations for the norms of X, X^*, X^{**} and X^{***}. If T is a continuous projection on X and $\|\cdot\|$ is the norm on it, $\|T\|$ denotes the corresponding norm of T. A subset M of X is said to be hypercompact if there is a sequence (x_n) in X with absolutely convex closed cover containing M and such that the sequence $(2^{pn}x_n)$ converges to the origin for every positive integer p.

We shall need after the following result that we have proved in Valdivia (1975):

a) *Let* X *and* Y *be two Banach spaces of infinite dimension with* Y^* *weak-star separable. If* A *is an absolutely convex and hypercompact subset of X, there is an absolutely convex and hypercompact subset* M of X which contains A such that X_M *is norm-isomorphic to* Y.

We denote by N the set of positive integers. Λ is the set of finite sequences of positive integers with an odd number of elements and strictly increasing, i.e., a member of Λ is $(r_1, r_2, \ldots, r_{2m+1})$, with r_j in N, j=1,2,..,2m+1, and

$$r_1 < r_2 < \ldots < r_{2m+1}.$$

Let us fix a vector space E. Let

$$U_1 \supset U_2 \supset \ldots \supset U_n \supset \ldots$$

be a sequence of absolutely convex subsets of E such that $L(U_n)$ is a reflexive Banach space with U_n as closed unit ball. We denote by E_n this Banach space and by $|\cdot|_n$ its norm. Let

$$U := \bigcap_{n=1}^{\infty} U_n.$$

We denote by F the Banach space L(U) with unit ball U. Let $|\cdot|$ be the norm of F. It is clear that

$$\lim_n |x|_n = |x|$$

when x belongs to F.

Let H be the vector space of all the sequences (x_n) with

$$x_n \in E_n \ , \ n=1,2,\ldots$$

If $x = (x_n)$ is in H, we write

$$\|x\| = \sup \left\{ \left[\sum_{j=1}^{m} |x_{r_{2j-1}} - x_{r_{2j}}|^2_{r_{2j-1}} + |x_{r_{2m+1}}|^2_{r_{2m+1}} \right]^{1/2} : (r_1, r_2, \ldots, r_{2m+1}) \in \Lambda \right\}$$

$$G := \{x \in H: \ \|x\| < \infty \}$$

It is obvious that $\|\cdot\|$ is a norm on G. In what follows we suppose G endowed with this norm. It is not difficult to prove that G is a Banach space.

Proposition 1. *If* $x = (x_n)$ *belongs to* G, *there exists an element* x_0 *in* F *such that*

$$\lim_n |x_n - x_0|_n = 0$$

Proof. Let us suppose that for some $\delta > 0$ and some sequence of integers

$$0 < r_1 < r_2 < \ldots < r_n < \ldots$$

we have

$$|x_{r_{2j-1}} - x_{r_{2j}}|_{r_{2j-1}} > \delta \ , \ j = 1,2,\ldots$$

Given a positive integer m, it follows that

$$\|x\|^2 > \sum_{j=1}^{m} |x_{r_{2j-1}} - x_{r_{2j}}|^2 > m\,\delta^2$$

and so $\|x\| = \infty$ which is a contradiction. Therefore, for every $\varepsilon > 0$ there is an integer n_0 such that

(1) $$|x_n - x_m|_n < \varepsilon \quad \text{when } m > n > n_0 .$$

As a consequence, (x_n) is a Cauchy sequence in E_r, $r = 1,2,\ldots$, and so it converges to some element x_0 in E_r, $r = 1,2,\ldots$ Obviously, x_0 belongs to $\bigcap_{r=1}^{\infty} E_r$ and it follows from (1) that

(2) $$|x_n - x_0|_n \le \varepsilon , \quad n > n_0.$$

Let us suppose that x_0 is not in F. If h is some number $h > \|x\|$ we know that x_0 does not belong to 2 h U and there is a positive integer s such that

$$x_0 \notin 2\, h\, U_s, \ |x_s - x_0|_s < h$$

after the inequality (2). Then

$$2\, h < |x_0|_s \le |x_s - x_0|_s + |x_s|_s \le h + \|x\| < 2\, h$$

which is a contradiction and the proof is finished.

We write L to denote the subspace of G formed by all the sequences (x_n) such that

$$\lim_n |x_n|_n = 0.$$

Let M be the subspace of G with elements (x_n) verifying

$$x_1 = x_2 = \ldots = x_n = \ldots$$

<u>Note 1</u>. If we take E = K and

$$U_n = \{x \in K: \ |x| \leq 1 \}, \ n=1,2,...$$

in the former construction, the space L obtained is nothing else that the space J of R.C. James (1950); see also Singer (1970), p. 273-178).

Proposition 2. *If* β *is the mapping from* F *into* M *defined by*

$$\beta x_0 = (x_0, x_0, ..., x_0, ...), \ x_0 \in F,$$

then β *is a norm-isomorphism from* F *onto* M.

Proof. It is immediate that β is linear, one to one and onto. If x_0 belongs to F and we write $x_n = x_0$, $n = 1,2,...$, it follows that

$$\|\beta x_0\| = \sup \left\{ \left(\sum_{j=1}^{m} |x_{r_{2j-1}} - x_{r_{2j}}|^2_{r_{2j-1}} + |x_{r_{2m+1}}|^2_{r_{2m+1}} \right)^{1/2} : \right.$$

$$\left. : (r_1, r_2, ..., r_{2m+1}) \in \Lambda \right\} =$$

$$= \sup \left\{ |x_0|_{r_{2m+1}} : r_{2m+1} \in N \right\} = \lim_n |x_0|_n = |x_0|.$$

q.e.d.

Proposition 3. G *is the topological direct sum of* L *and* M.

Proof. M is closed in G because it is norm-isomorphic to F. It is not difficult to show that L also is closed in G and $M \cap L = \{0\}$. On the other hand, if $x=(x_n)$ belongs to G, by Proposition 1, we can find a vector x_0 in F such that

$$\lim_n |x_n - x_0|_n = 0.$$

Let $y = (y_n)$, $z = (z_n)$, be defined by $y_n = x_n - x_0$, $z_n = x_0$, n= 1,2,... Then

$$y \in L, \ z \in M, \ x = y + z.$$

q.e.d.

Proposition 4. *We have*

$$\|I - T\| \leq 1, \quad \|T\| \leq 1.$$

Proof. Let us take any element $x = (x_n)$ in G. Let x_0 be the point in F such that

$$\lim_n |x_n - x_0|_n = 0.$$

We write $y = (y_n)$, $z = (z_n)$, with $y_n = x_n - x_0$, $z_n = x_0$, $n = 1,2,\ldots$ Then, if we apply Proposition 2, we obtain:

$$\|(I - T)\, x\| = \|z\| = |x_0| .$$

Given any $\varepsilon > 0$, we can find a positive integer s such that

$$|x_0| < |x_0|_s + \varepsilon , \quad |x_s - x_0|_s < \varepsilon.$$

Then

$$\|(I - T)x\| = |x_0| \leq |x_0 - x_s|_s + |x_s|_s + \varepsilon \leq |x_s|_s + 2\varepsilon \leq$$
$$\leq \|x\| + 2\varepsilon$$

and thus

$$\|I - T\| \leq 1.$$

Given any $\varepsilon > 0$, we can find $(n_1, n_2, \ldots, n_{2r+1})$ in Λ and integer $n_{2r+2} > n_{2r+1}$ such that

$$\|Tx\| = \|y\| < \left[\sum_{j=1}^{r} |y_{n_{2j-1}} - y_{n_{2j}}|^2_{n_{2j-1}} + |y_{n_{2r+1}}|^2_{n_{2r+1}}\right]^{1/2} + \frac{1}{2}\,\varepsilon$$

$$|y_{n_{2r+2}}|_{n_{2r+1}} < \frac{1}{2}\,\varepsilon$$

Then

$$\|Tx\| = \|y\| < \left[\sum_{j=1}^{r} |y_{n_{2j-1}} - y_{n_{2j}}|^2_{n_{2j-1}} + |y_{n_{2r+1}}|^2_{n_{2r+1}} \right]^{1/2} + \frac{1}{2}\varepsilon$$

$$\leq \sum_{j=1}^{r} |y_{n_{2j-1}} - y_{n_{2j}}|^2_{n_{2j-1}} + |y_{n_{2r+1}} - y_{n_{2r+2}}|^2_{n_{2r+1}} \Big]^{1/2} +$$

$$+ |y_{n_{2r+2}}|_{n_{2r}} + \frac{1}{2}\varepsilon$$

$$\leq \left[\sum_{j=1}^{r} |x_{n_{2j-1}} - x_{n_{2j}}|^2_{n_{2j-1}} + |x_{n_{2r+1}} - x_{n_{2r+2}}|^2_{n_{2r+1}} \right]^{1/2} +$$

$$+ |y_{n_{2r+2}}|_{n_{2r+2}} + \frac{1}{2}\varepsilon \leq \|x\| + \varepsilon ,$$

from where it follows that $\|T\| \leq 1$. q.e.d.

Let us write for every positive integer m:

$$L_m := \left\{ x\{m\} : x \in G \right\}.$$

It is immediate that L_m is norm-isomorphic to E_m and that the linear hull of

$$\cup \left\{ L_m : m = 1,2,\ldots \right\}$$

is dense in L.

Proposition 5. *(L_n) is a Schauder decomposition of L which is shrinking and monotone.*

Proof. Let (x_n^j) be an element of L_j, $j = 1,2,\ldots,p+1,\ldots,p+q$. We put

$$u := \sum_{j=1}^{p} (x_n^j), \quad v := \sum_{j=1}^{p+q} (x_n^j).$$

Then

$$\|u\| = \|v[p]\| \le \|v[p + q]\| = \|v\|$$

and therefore, (L_n) is a monotone Schauder decomposition of L. To show now that (L_n) is shrinking we shall apply an analogous method to that used in James (1950). Let us suppose that (L_n) is not shrinking. From [Singer (1981), pp. 524-525] we know the existence of f in L^*, an $\varepsilon > 0$, a sequence of positive integers

$$1 = n_1 < n_2 < \dots < n_q < \dots$$

and

$$x^j = (x^j_n) \in L_j, \; j = 1,2,\dots$$

such that when we write

$$z^q = \sum_{j=n_q}^{n_{q+1}-1} x^j$$

we have

$$\|z^q\| = 1, \; f(z^q) \ge \varepsilon, \; q = 1,2,\dots$$

Let (y_n) be the element of H defined by

$$y_n = \frac{1}{q} x^n_n, \; n_q \le n < n_{q+1}, \; q = 1,2,\dots$$

We take $(r_1, r_2, \dots, r_{2m+1})$ in Λ and fix a positive integer h. If there is a smallest positive integer p and a greatest integer q such that

$$n_h \le r_{2p-1} < r_{2q} < n_{h+1}$$

it follows that

$$\sum_{j=p}^{q} \left| y_{r_{2j-1}} - y_{r_{2j}} \right|_{r_{2j-1}}^{2} = \frac{1}{h^2} \sum_{j=p}^{q} \left| x_{r_{2j-1}}^{r_{2j-1}} - x_{r_{2j}}^{r_{2j}} \right|_{r_{2j-1}}^{2} \leq$$

$$\leq \frac{\|z^h\|^2}{h^2} = \frac{1}{h^2}$$

On the other hand, if given a positive integer j, with $1 \leq j \leq m$, there are two positive integers $h_1 < h_2$ such that

$$n_{h_1} \leq r_{2j-1} < n_{h_1+1} \quad , \quad n_{h_2} \leq r_{2j} < n_{h_2+1} \; ,$$

it follows that

$$\left| y_{r_{2j-1}} - y_{r_{2j}} \right|_{r_{2j-1}} = \left| \frac{1}{h_1} x_{r_{2j-1}}^{r_{2j-1}} - \frac{1}{h_2} x_{r_{2j}}^{r_{2j}} \right|_{r_{2j-1}} \leq$$

$$\leq \frac{1}{h_1} \left| x_{r_{2j-1}}^{r_{2j-1}} \right|_{r_{2j-1}} + \frac{1}{h_2} \left| x_{r_{2j}}^{r_{2j}} \right|_{r_{2j}} \leq \frac{1}{h_1} \|z^{h_1}\| + \frac{1}{h_2} \|z^{h_2}\| <$$

$$< \frac{2}{h_1}$$

and therefore

$$\sum_{j=1}^{m} \left| y_{r_{2j-1}} - y_{r_{2j}} \right|_{r_{2j-1}}^{2} + \left| y_{r_{2m+1}} \right|_{r_{2m+1}}^{2} \leq \sum_{n=1}^{\infty} \frac{1}{n^2} + \sum_{n=1}^{\infty} \frac{4}{n^2} + 1$$

from where (y_n) belongs to G. Moreover

$$\lim_{n} |y_n|_n = 0$$

and (y_n) even belongs to L. Let us finally observe that

$$f((y_n)) = f\left(\sum_{q=1}^{\infty} \frac{1}{q} z^q\right) \geq \varepsilon \sum_{q=1}^{\infty} \frac{1}{q} = \infty,$$

which is the contradiction we are looking for.

q.e.d.

Let A be the closed unit ball of G and $B := A \cap L$. We denote by S_m the subspace of L^* which is orthogonal to the linear hull of

$$\bigcup \{L_n : n \neq m,\ n \in N\}$$

Then (S_m) is a Schauder decomposition of L^*. Given any element u of L^*, we have

$$u = \sum_{n=1}^{\infty} u_n,\ u_n \in S_n,\ n = 1,2,\ldots$$

For every positive integer n we can identify u_n with the element $\psi_n(u_n)$ in E_n^* such that

$$u_n(x) = \psi_n(u_n)(x_n)$$

for any $x = (x_n)$ in L_n. Let us write

$$\psi(u) = (\psi_n(u_n))$$

So we can consider the elements of L^* as sequences $v = (v_n)$ defined through the mapping ψ with

$$v_n \in E_n^*,\ n = 1,2,\ldots,\ v(y) = \sum_{n=1}^{\infty} \langle y_n, v_n\rangle,\ y = (y_n) \in L.$$

Then, for every positive integer m,

$$S_m = \{v \in L^* : v = v\{m\}\}$$

Let us now take

$$y = (y_n) \in G, \; v = (v_n) \in L^*.$$

If we bear in mind that $y[m]$ belongs to L, $m = 1,2,\ldots,$ and that the Schauder decomposition (L_n) of L is monotone it follows that

$$\left|\sum_{j=1}^{n} \langle y_j, v_j\rangle\right| = |\langle y|n|, v\rangle| \le \|y[n]\| \cdot \|v\| \le \|y\| \cdot \|v\|$$

and so

$$|\langle y, v\rangle| = \left|\sum_{j=1}^{\infty} \langle y_j, v_j\rangle\right| = \lim_n \left|\sum_{j=1}^{n} \langle y_j, v_j\rangle\right| \le \|y\|\cdot\|v\|$$

which allows us to look at the elements of G as vectors in L^{**} and A is contained in B too.

Let us fix any element φ in L^{**}. If $v=(v_n)$ belongs to L^* we write

$$\varphi_n(v_n) = \varphi(v\{n\}), \; n = 1,2,\ldots$$

Since S_n is reflexive it follows that φ_n belongs to E_n and so $\lambda = (\varphi_n)$ is an element of H. Obviously

$$\varphi(v) = \langle\lambda, v\rangle = \sum_{n=1}^{\infty} \langle\varphi_n, v_n\rangle .$$

Let us take any element $(r_1, r_2, \ldots, r_{2m+1})$ in Λ. Let h be an integer greater than r_{2m+1}. Then, since $\lambda[h]$ belongs to L:

(3) $$\sum_{j=1}^{m} |\varphi_{r_{2j-1}} - \varphi_{r_{2j}}|^2_{r_{2j-1}} + |\varphi_{r_{2m+1}}|^2_{r_{2m+1}} \leq \|\lambda\,[h]\|^2$$

We find an element $w = (w_n)$ in L^*, $\|w\| = 1$ such that

$$|\langle\lambda[h], w\rangle| = \|\lambda[h]\|.$$

Using that (S_n) is a monotone Schauder decomposition we can see that

$$\|\lambda[h]\| = |\langle\lambda[h], w\rangle| = |\langle\lambda, w[h]\rangle| = |\varphi(w[h])| \leq \|\varphi\| \cdot \|w[h]\| \leq \|\varphi\|$$

and from (3) we see that λ belongs to G and

$$\|\lambda\| \leq \|\varphi\|.$$

Thus, if we identify φ with λ it follows that A contains B. Therefore A=B.

It follows from the former identification that G is the second conjugate Banach space of L. In what follows we shall write $G = L^{**}$.

We also write

$$S := \left\{ u \in L^* : \langle x, u\rangle = 0 : x \in M \right\}.$$

Proposition 6. *M is the subspace of* L^{**} *which is orthogonal to* S.

Proof. Let

$$\left\{ y^c = (y^c_n) : c \in C , \geq \right\}$$

be a net in M which converges to $y = (y_n)$ in L^{**} for the weak-star topology. Suppose that there exist two positive integers $h < k$ such that $y_h \neq y_k$. Then, for some w in E^*_h ,

$$\langle y_h, w\rangle \neq \langle y_k, w\rangle.$$

Let $u = (u_n)$ be the element of S such that $u_h = -u_k = w$ and $u_n = 0$ for every positive integer n different from h and k. We have

$$0 \neq \langle y_h - y_k, w\rangle = \langle y, u\rangle = \lim_c \left\{ \langle y^c, u\rangle : c \in C, \geq \right\} = 0$$

Which is a contradiction.

q.e.d.

Proposition 7. *If we identify every element of* L *with its restriction on* S *then* $S^* = L$.

Proof. The mapping ζ that assignes to every x in L its restriction on S obviously is linear and one to one. On the other hand if z belongs to S^* an application of the Hahn-Banach theorem gives us an element y in L^{**} such that y restricted to S is nothing else that z. We have

$$y = y^1 + y^2, \quad y^1 \in L, \; y^2 \in M.$$

Then $\zeta(y^1) = L$. Therefore

$$\zeta : L \longrightarrow S^*$$

is a bijection. We can identify L with S^* through the mapping ζ. To finish we see how B is $\sigma(L,S)$-closed. We take any element $x = (x_n)$ in L which is in the polar set of $B^0 \cap S$ in L. Let

$$\left\{ x^c = (x^c_n) : c \in C, \geq \right\} \tag{4}$$

be a net in B such that for every v in S we have

$$\lim_c \left\{ \langle x^c, v\rangle : c \in C, \geq \right\} = \langle x, v\rangle.$$

The net (4) has a cluster point $y = (y_n)$ in L^{**} for the weak-star topology such that y belongs to $A = \tilde{B}$. Then, according to Proposition 4,

$$\|Ty\| \le \|y\| \le 1$$

and therefore Ty belongs to B. Ty and x coincide on S and so $Ty = x$.

q.e.d.

Given a positive integer n, let $\mathfrak{T}$ be the topology on U which is the restriction of the weak topology on E_n. Since that space is reflexive $\mathfrak{T}$ does not depend on the positive integer n. Let us denote by P the vector space of all linear forms on F whose restriction to U is $\mathfrak{T}$-continuous. If W is the polar set of U in P and we suppose that P is endowed with the norm given by the gauge of W, then P is a Banach space and its dual can be identified in the obvious way with F.

<u>Proposition 8</u>. L^*/S *is norm-isomorphic to* P.

<u>Proof</u>. We apply Proposition 2 and obtain that β transforms U in $\tilde{B} \cap M$. It is not difficult to see that

$$\beta:\ F[\sigma(F,P)] \longrightarrow M[\sigma(M,L^* \mid S)]$$

is an isomorphism and so the mapping from L^*/S onto P adjoint of β is a norm-isomorphism.

q.e.d.

Let Γ be the set of all increasing finite-sequences of positive integers with two terms at least, i.e., $(r_1, r_2, \ldots, r_{m+1})$ belongs to Γ if r_j is in N, $j=1,2,\ldots,m+1$, $m \ge 1$ and

$$r_1 < r_2 < \ldots < r_{m+1}$$

For every $x = (x_n)$ in L, we write

$$\||x|\| = \frac{1}{\sqrt{2}} \sup \left\{ \left(\sum_{j=1}^{m} |x_{r_j} - x_{r_{j+1}}|^2_{r_j} + |x_{r_1} - x_{r_{m+1}}|^2_{r_1} \right)^{\frac{1}{2}} :\right.$$

$$:(r_1, r_2, \ldots, r_{m+1}) \in \Gamma \Big\} .$$

It can be proved that $|||\cdot|||$ is a norm on L, which is equivalent to $\|\cdot\|$. (L_n) is a monotone Schauder decomposition of $(L, |||\cdot|||)$. We set

$$D := \Big\{ x \in L : |||x||| \le 1 \Big\} .$$

Note 2. If E = K and

$$U_n = \Big\{ x \in K : |x| \le 1 \Big\} , n = 1,2,\ldots,$$

then $\sqrt{2}\ |||\cdot|||$ is the norm used in James (1951) to show that $(J, \sqrt{2}|||\cdot|||)^{**}$ is norm-isomprohic to $(J, \sqrt{2}\ |||\cdot|||)$.

Proposition 9. *If* $x = (x_n)$ *belongs to* L^{**}, *then*

$$|||x||| = \lim_r |||\ x[r]\ |||.$$

Proof. It is enough to prove the case $|||x||| = 1$. Let us fix a positive integer j. Since

$$D^0 \cap (S_1 + S_2 + \ldots + S_j)$$

is a weakly compact subset of L^* there is an element $y^j = (y^j_n)$ in D such that

$$|\ \langle x - y^j, u\rangle\ | < \frac{1}{j} , \quad u \in D^0 \cap (S_1 + S_2 + \ldots + S_j).$$

As the linear hull of

$$\cup \Big\{ D^0 \cap (S_1 + S_2 + \ldots + S_j) : j = 1,2,\ldots \Big\}$$

is dense in L^*, it follows that (y^j) converges to x for the weak-star topology of L^{**}. For every positive integer n, (y^j_n) converges to x_n in E_n. Consequently, the sequence $(y^j_n[r])$ converges to $x[r]$ in L for every fixed positive integer r. Therefore

$$\| |x[r]| \| \leq 1.$$

The sequence $(\| |x[r]| \|)$ is increasing and has a limit α less than, or equal to one. On the other hand, $(x[r])$ belongs to αD and converges to x in the weak-star topology of L^{**}, so x belongs to $\alpha\tilde{D}$ and $\alpha = 1$ because $\| |x| \| = 1$.

q.e.d.

Proposition 10. *D is $\sigma(L,S)$-closed.*

Proof. Given the positive integers $p < q$ and the vector u of E^*_p, we denote by $u(p,q)$ the vector (v_n) of L^* such that

$$v_p = - v_q = u, \quad v_n = 0 \text{ if } n \neq p \text{ and } n \neq q.$$

Let us denote

$$A(p,q) = \left\{ u(p,q) \ : \ u \in E^*_p , \ |u|_p \leq 1 \right\} \qquad \text{and}$$

$$B_j := \cup \left\{ A(p,q) : \ p+q \leq j+1 \right\}, \ j = 1,2,\ldots$$

Take a vector $x = (x_n)$ in the $\sigma(L,S)$-closure of D. For every positive integer j, B_j is weakly compact in S and we can find an element $y^j = (y^j_n)$ in D such that

$$|\langle x-y^j, w\rangle| < \frac{1}{j}, \ w \in B_j,$$

since D is a convex set. Given $\varepsilon > 0$, we find an element $(r_1, r_2, \ldots, r_{m+1})$ in Γ such that

$$\| |x| \| - \varepsilon < 1/\sqrt{2} \left(\sum_{j=1}^{m} |x_{r_j} - x_{r_{j+1}}|_{r_j}^2 + |x_{r_1} - x_{r_{m+1}}|_{r_1}^2 \right)^{1/2}$$

Let us take an integer

$$j_0 > 2r_{m+1}, \quad \frac{1}{j_0} < \frac{\varepsilon}{m+1}.$$

We find

$$u_j \in E^*_{r_j}, \quad |u_j|_{r_j} = 1, \quad j=1,2,\ldots,m, \quad u_{m+1} \in E^*_{r_1}, \quad |u_{m+1}|_{r_1} = 1,$$

with

$$|x_{r_j} - x_{r_{j+1}}|_{r_j} = \langle x_{r_j} - x_{r_{j+1}}, u_j \rangle, \quad j=1,2,\ldots,m,$$

$$|x_{r_1} - x_{r_{m+1}}|_{r_1} = \langle x_{r_1} - x_{r_{m+1}}, u_{m+1} \rangle$$

Therefore,

$$|x_{r_j} - x_{r_{j+1}}|_{r_j} = |\langle x_{r_j} - x_{r_{j+1}} - (y^{j_0}_{r_j} - y^{j_0}_{r_{j+1}}), u_j \rangle| +$$

$$+ |\langle y^{j_0}_{r_j} - y^{j_0}_{r_{j+1}}, u_j \rangle| \le$$

$$\le |\langle x - y^{j_0}, u_j(r_j, r_{j+1}) \rangle| + |y^{j_0}_{r_j} - y^{j_0}_{r_{j+1}}|_{r_j} \le$$

$$\le \frac{1}{j_0} + |y^{j_0}_{r_j} - y^{j_0}_{r_{j+1}}|_{r_j} <$$

$$< \frac{\varepsilon}{m+1} + |y^{j_0}_{r_j} - y^{j_0}_{r_{j+1}}|_{r_j}, \quad j = 1,2,\ldots,m.$$

Analogously,

$$|x_{r_1} - x_{r_{m+1}}|_{r_1} < \frac{\varepsilon}{m+1} + |y^{j_0}_{r_1} - y^{j_0}_{r_{m+1}}|_{r_1}$$

Then

$$\| |x| \| - \varepsilon < \frac{1}{\sqrt{2}} \left(\sum_{j=1}^{m} \left(\frac{\varepsilon}{m+1} + |y_{r_j}^{j_0} - y_{r_{j+1}}^{j_0}|_{r_j} \right)^2 + \right.$$

$$\left. + \left(\frac{\varepsilon}{m+1} + |y_{r_1}^{j_0} - y_{r_{m+1}}^{j_0}|_{r_1} \right)^2 \right)^{1/2} \le$$

$$\le \frac{1}{\sqrt{2}} \left[\sum_{j=1}^{m+1} \frac{\varepsilon^2}{(m+1)^2} \right]^{1/2} + \frac{1}{\sqrt{2}} \left[\sum_{j=1}^{m} |y_{r_j}^{j_0} - y_{r_{j+1}}^{j_0}|_{r_j}^2 + |y_{r_1}^{j_0} - y_{r_{j+1}}^{j_0}|_{r_1}^2 \right]^{1/2} <$$

$$< \varepsilon + \| |y^{j_0}| \| \le \varepsilon + 1;$$

So it follows that $\| |x| \| \le 1$.

q.e.d.

Proposition 11. *We have*

$$\| |I - T| \| \le 1 \ , \quad \| |T| \| \le 1.$$

Proof. $\| |T| \|$ is less than or equal to one because $(L^{**}, \| |\cdot| \|)$ is the third conjugate of $(S, \| |\cdot| \|)$. Let us now take an element $x = (x_n)$ in L^{**}. If

$$z = (z_n) := (I - T)x, \text{ with } z_n = x_0 , \ n = 1,2,\ldots,$$

we know after Proposition 1 that

$$\lim_n |x_n - x_0|_n = 0$$

and so

$$\lim_n |x_n|_n = \lim_n |x_0|_n = |x_0|.$$

Given a positive integer r, we set $x[r] = (y_n)$. Then, since $(r, r+1)$

belongs to Γ, we have

$$\| |x[r]| \| \geq \frac{1}{\sqrt{2}} \left(|y_r - y_{r+1}|_r^2 + |y_r - y_{r+1}|_r^2 \right)^{1/2} = |y_r|_r =$$

$$= |x_r|_r$$

and therefore, bearing in mind Proposition 9,

$$|x_0| = \lim_r |x_r|_r = \lim_r \| |x[r]| \| = \| |x| \|.$$

On the other hand,

$$\| |(I - T)x| \| = \lim_r \| |z[r]| \| = \lim_r |z_r|_r = |x_0|.$$

Thus $\quad \| |(I - T)x| \| \leq \| |x| \|.$

q.e.d.

Proposition 12. $(L^*, \| |\cdot| \|)/S$ *is norm-isomorphic to* P.

Proof. If we realize that β transforms U in $\tilde{D} \cap M$ the proof is the same that in Proposition 8.

q.e.d.

Proposition 13. *If* $U_1 = U_2 = \ldots = U_n = \ldots$ *then* $(L^*, \| |\cdot| \|)$ *is norm-isomorphic to* $(L^{**}, \| |\cdot| \|)$.

Proof. Let ξ be the mapping from L^* into L^{**} such that, if $x = (x_n)$ belongs to L^* then $\xi(x) = y = (y_n)$ with

$$y_n = x_{n+1} - x_n, \quad n = 1,2,\ldots$$

Obviously, ξ is a linear bijection. Given $(r_1, r_2, \ldots, r_{m+1})$ in Γ, we have

$$\sum_{j=1}^{m} |y_{r_j} - y_{r_{j+1}}|_{r_j}^2 + |y_{r_1} - y_{r_{m+1}}|_{r_1}^2 = \sum_{j=1}^{m} |x_{r_j+1} - x_{r_{j+1}+1}|_{r_j+1}^2 +$$

$$+ |x_{r_1+1} - x_{r_{m+1}+1}|^2_{r_1+1} \leq \||x|\|^2$$

and therefore

$$(5) \qquad \||\xi(x)|\| \leq \||x|\| .$$

On the other hand, if $r_1 > 1$ we have

$$\sum_{j=1}^{m} |x_{r_j} - x_{r_{j+1}}|^2_{r_j} + |x_{r_1} - x_{r_{m+1}}|^2_{r_1} =$$

$$(6) \qquad \sum_{j=1}^{m} |y_{r_j-1} - y_{r_{j+1}-1}|^2_{r_j-1} + |y_{r_1-1} - y_{r_{m+1}-1}|^2_{r_1-1} \leq \|\xi(x)\|^2 ,$$

and for $r_1=1$, we take integers r and r_{m+2} such that

$$r_{m+1} < r < r_{m+2}-1 ,$$

write $y[r] = (z_n)$ and compute

$$\sum_{j=1}^{m} |x_{r_j} - x_{r_{j+1}}|^2_{r_j} + |x_{r_1} - x_{r_{m+1}}|^2_{r_1} =$$

$$\sum_{j=2}^{m} |x_{r_j} - x_{r_{j+1}}|^2_{r_j+1} + |x_{r_1} - x_{r_2}|^2_{r_1} + |x_{r_1} - x_{r_{m+1}}|^2_{r_1} =$$

$$= \sum_{j=2}^{m} |y_{r_j-1} - y_{r_{j+1}-1}|^2_{r_j-1} + |y_{r_2-1}|^2_{r_1} + |y_{r_{m+1}-1}|^2_1 =$$

$$= \sum_{j=2}^{m} |z_{r_j-1} - z_{r_{j+1}-1}|^2_{r_j-1} + |z_{r_{m+1}-1} - z_{r_{m+2}-1}|^2_{r_{m+1}-1} +$$

$$(7) \qquad + |z_{r_2-1} - z_{r_{m+2}-1}|^2_{r_2-1} \leq \||y[r]|\|^2 \leq \||\xi(x)|\|^2$$

It follows now from (5), (6) and (7) that

$$\| |\xi(x)| \| = \| |x| \|. \qquad \text{q.e.d.}$$

Let us write for any positive integer m

$$M_n := \left\{ x = (x_n) \in L : x = x[m] \; , \; x_1 = x_2 = \ldots = x_m \right\}.$$

Let η_m be the mapping from M_m into L_m defined by $\eta_m(x) = x\{m\}$ if $x=(x_n)$ belongs to M_m. We have that

$$\eta_m : (L_m \, , \; \|\cdot\|) \longrightarrow (M_m \, , \; \|\cdot\| \;)$$

and

$$\eta_m : (L_m \, , \; \||\cdot|\|) \longrightarrow (M_n \, , \; \||\cdot|\|)$$

are norm-isomorphisms.

<u>Proposition 14</u>. (M_n) *is a Schauder decomposition of* L *which is monotone in* $(L \, , \; \|\cdot\|)$ *and* $(L, \; \||\cdot|\|)$.

<u>Proof</u>. The linear hull of

$$\cup \left\{ M_n \; : \; n = 1,2,\ldots \right\}$$

is obviously dense in L. Let

$$x^j = (x_n^j) \in M_j \; , \; j = 1,2,\ldots,p,p+1,\ldots,p+q.$$

If we write

$$y = (y_n) = \sum_{j=1}^{p} x^j \,, \quad z = (z_n) = \sum_{j=1}^{p+q} x^j \,,$$

then we have

$$y_n = \sum_{j=n}^{p} x_n^j \,, \quad n = 1,2,\ldots,p \,, \quad y_n = 0 \,, \quad n = p+1, p+2, \ldots$$

$$z_n = \sum_{j=n}^{p+q} x_n^j \,, \quad n = 1,2,\ldots,p,p+1,\ldots p+q, \quad z_n=0, \quad n= p+q+1, p+q+2, \ldots$$

So, it follows that

$$\|u\| \le \|v\|, \quad \||u|\| \le \||v|\|$$

and thus, M_n is a monotone Schauder decomposition in $(L, \|\cdot\|)$ and $(L, \||\cdot|\|)$.

q.e.d.

If m is any positive integer, we write

$$T_m := \left\{ u = (u_n) \in L^* \; : \; u = u\{m\} + u\{m+1\} \,, \; u_m = - u_{m+1} \right\}.$$

It is easy to see that T_m is the subspace of L^* which is orthogonal to

$$\cup \left\{ M_n \; : \; n = 1,2,\ldots,m-1,m+1,\ldots \right\}.$$

and that the closed linear hull of

$$\cup \left\{ T_n \; : \; n = 1,2,\ldots \right\}$$

in L^* coincides with S. Therefore, (T_n) is a monotone Schauder decomposition of $(S, \|\cdot\|)$ and $(S, \||\cdot|\|)$.

Each one of $(S_n, \|\cdot\|)$, $(T_n, \|\cdot\|)$, $(S_n, |||\cdot|||)$ and $(T_n, |||\cdot|||)$ is norm-isomorphic to L_n^*, $n = 1,2,\ldots$

2. BANACH SPACES Z^{**}/Z.

Following the research initiated in James (1960), Lindenstrauss (1971) shows that when X is a separable Banach space, there is a Banach space Z such that Z^{**}/Z is isomorphic to X and Z^* has a shrinking Schauder basis. Davis et al. (1974) prove that when X is a weakly compactly generated Banach space there is a Banach space Z such that Z^{**}/Z is isomorphic to X and Z^* has a Schauder decomposition (Y_n) where Y_n is isomorphic to a certain reflexive Banach space with unconditional basis. We prove now results related with the former ones using norm-isomorphism instead of isomorphism.

Theorem 1. *Let X be a separable space. Let (X_n) be a sequence of infinite dimensional, separable and reflexive Banach spaces. Then there is a Banach space Z with the following properties:*

1) Z^{**}/Z is norm-isomorphic to X.

2) Z has a Schauder decomposition (Z_n) which is monotone and shrinking with Z_n isomorphic to X_n, $n = 1,2,\ldots$

3) Z^* has a Schauder decomposition (Y_n) which is monotone and shrinking with Y_n isomorphic to X_n^*, $n = 1,2,\ldots$

4) If $Z^{\perp}$ is the subspace orthogonal to Z in Z^{***}, the projection of Z^{***} onto $Z^{\perp}$ has norm one.

5) Z^{***}/Z^* is norm-isomorphic to X^*.

Proof. Let us take an infinite dimensional separable Banach space Y and denote by V its unit ball. We find a sequence (y_n) in Y with dense linear hull and such that $(2^{pn} y_n)$ converges to the origin for every positive integer p. We can sucessively apply result a) to obtain an increasing sequence (P_n) of hyperprecompact subsets of Y such that

$$2^{pn} y_n \in P_1$$

and X_{P_n} is norm-isomorphic to X_n, $n = 1,2,\ldots$ We denote by E the algebraic dual of Y. Let V_n be the polar set of P_n in E, $n = 1,2,\ldots$ Let V_0 be the polar set of V in E. Let us suppose that $L(V_0)$ is endowed with the norm of unit ball V_0.

If X is infinite dimensional, we take Y = X and denote by U the set V_0. If X has finite dimension, we take in $L(V_0)$ an absolutely convex and compact subset U such that $L(V_0)_U$ is norm-isomorphic to X.

Let us write

$$U_n := U + \frac{1}{n} V_n, \quad n = 1,2,\ldots$$

Then, the spaces L_n, M_n and T_n constructed in the former section are isomorphic to X_n^*, X_n^* and X_n respectively, $n = 1,2,\ldots$ and

$$U = \bigcap_{n=1}^{\infty} U_n.$$

We take Z as the space S constructed there. Then 1), 2) 3) and 4) follow from propositions 8, 14, 5 and 4, respectively. Result 5) follows from 4)

q.e.d.

Theorem 2. *Let X be a weakly compactly generated Banach space. Then there exist a reflexive Banach space Q with unconditional basis and a Banach space Z verifying the following proporties:*

1) Z^{**}/Z is norm-isomorphic to X.

2) Z has a Schauder decomposition (Z_n), which is monotone and shrinking with Z_n isomorphic to Q, $n = 1,2,\ldots$

3) Z^* has a Schauder decomposition (Y_n), which is monotone and shrinking with Y_n isomorphic to Q^*, $n = 1,2,\ldots$

4) If $Z^{\perp}$ is the orthogonal subspace of Z in Z^{***}, the projection of Z^{***} onto $Z^{\perp}$ along Z^* has norm one.

5) Z^{***}/Z^* is norm-isomorphic to X^*.

Proof. After the former theorem we can reduce ourselves to the nonseparable case. Let V be the closed unit ball of X. We can find an absolutely convex and weakly compact W of X such that X_W is reflexive with unconditional basis and dense in X , [1]. Let Q be a Banach space isomorphic to X_W. We set E to denote the algebraic dual of X; U and W_0 are the polar sets of V and W in E, respectively. Let us write

$$U_n := U + \frac{1}{n} W_0 \, , \ n = 1,2,\ldots$$

The proof finishes as in the former theorem. q.e.d.

Theorem 3. *If X is a nonzero reflexive Banach space, there is a Banach space Z with the following properties:*

1) Z^{**}/Z is norm-isomorphic to X.

2) There is a monotone and shrinking Schauder decomposition (Z_n) in Z such that Z_n is norm-isomorphic to X, n = 1,2,...

3) There is a monotone and shrinking Schauder decomposition (Y_n) in Z^* such that Y_n is norm-isomorphic to X^*, n =1,2,...

4) If $Z^{\perp}$ is the orthogonal subspace of Z in Z^{***}, the projection of Z^{***} onto $Z^{\perp}$ along Z^* has norm one.

5) Z^{***}/Z^* is norm-isomorphic to X^*.

6) Z^{***} is norm-isomorphism to Z^*.

Proof. Let V be the closed unit ball of X. We set U for the polar set of V in X^*. We take $E = X^*$, $U_n = U$, n = 1,2,... Then, the spaces $(L_n, \|\!|\cdot|\!\|)$, $(M_n, \|\!|\cdot|\!\|)$ and $(T_n, \|\!|\cdot|\!\|)$ constructed in the former section are norm-isomorphic to X^*, X^* and X, respectively. We take Z as the space $(S, \|\!|\cdot|\!\|)$ constructed there. Then 1), 2) 3) and 4) follow from proposition 12, 14 , 5 and 11, respectively. Property 5) follows from 4) , and 6) is obtained from Proposition 11. q.e.d.

BIBLIOGRAPHY.

Davis, W. J. ; Figiel, T. ; Johnson, W.B. & Pelczynski, A. (1974). Factoring weakly compact operators, J. Funct. Anal. 17, 311-327.

James, R.C. (1950). Bases and reflexivity of Banach spaces, Ann. of Math. 52, 518-527.

James, R. C. (1951). A non-reflexive Banach space isometric with its second conjugate, Proc. Nat. Acad. Sci. (U.S.A.) 37, 174-177.

James, R.C. (1960). Separable conjugate spaces, Pacific J. Math. 10, 563-571.

Lindenstrauss, J. (1971). On James' paper "separable conjugate spaces", Israel J. Math. 9, 279-284.

Singer, I. (1970). Bases in Banach spaces I. Berlin-Heidelberg-New York, Springer.

Singer, I. (1981). Bases in Banach spaces II. Berlin-Heidelberg-New York, Springer.

Valdivia, M. (1975). A class of precompact sets in Banach spaces, J. Reine Angew Math. 276, 130-135.